P9-EEB-738

ROOFS AND SIDING

Other Publications:

LIBRARY OF HEALTH
CLASSICS OF THE OLD WEST
THE EPIC OF FLIGHT
THE GOOD COOK
THE SEAFARERS
THE ENCYCLOPEDIA OF COLLECTIBLES
THE GREAT CITIES
WORLD WAR II
THE WORLD'S WILD PLACES
THE TIME-LIFE LIBRARY OF BOATING
HUMAN BEHAVIOR
THE ART OF SEWING
THE OLD WEST
THE EMERGENCE OF MAN
THE AMERICAN WILDERNESS
THE TIME-LIFE ENCYCLOPEDIA OF GARDENING
LIFE LIBRARY OF PHOTOGRAPHY
THIS FABULOUS CENTURY
FOODS OF THE WORLD
TIME-LIFE LIBRARY OF AMERICA
TIME-LIFE LIBRARY OF ART
GREAT AGES OF MAN
LIFE SCIENCE LIBRARY
THE LIFE HISTORY OF THE UNITED STATES
TIME READING PROGRAM
LIFE NATURE LIBRARY
LIFE WORLD LIBRARY
FAMILY LIBRARY:
   HOW THINGS WORK IN YOUR HOME
   THE TIME-LIFE BOOK OF THE FAMILY CAR
   THE TIME-LIFE FAMILY LEGAL GUIDE
   THE TIME-LIFE BOOK OF FAMILY FINANCE

HOME REPAIR
AND IMPROVEMENT

ROOFS AND SIDING

BY THE EDITORS OF
TIME-LIFE BOOKS

TIME-LIFE BOOKS
ALEXANDRIA, VIRGINIA

Time-Life Books Inc.
is a wholly owned subsidiary of
TIME INCORPORATED

Founder — Henry R. Luce 1898-1967

Editor-in-Chief — Henry Anatole Grunwald
President — J. Richard Munro
Chairman of the Board — Ralph P. Davidson
Executive Vice President — Clifford J. Grum
Chairman, Executive Committee — James R. Shepley
Editorial Director — Ralph Graves
Group Vice President, Books — Joan D. Manley
Vice Chairman — Arthur Temple

TIME-LIFE BOOKS INC.

Managing Editor — Jerry Korn
Executive Editor — David Maness
Assistant Managing Editors — Dale M. Brown (planning), George Constable, Thomas H. Flaherty Jr. (acting), Martin Mann, John Paul Porter

Art Director — Tom Suzuki
Chief of Research — David L. Harrison
Director of Photography — Robert G. Mason
Assistant Art Director — Arnold C. Holeywell
Assistant Chief of Research — Carolyn L. Sackett
Assistant Director of Photography — Dolores A. Littles

Chairman — John D. McSweeney
President — Carl G. Jaeger
Executive Vice Presidents — John Steven Maxwell, David J. Walsh
Vice Presidents — George Artandi (comptroller); Stephen L. Bair (legal counsel); Peter G. Barnes; Nicholas Benton (public relations); John L. Canova; Beatrice T. Dobie (personnel); Carol Flaumenhaft (consumer affairs); James L. Mercer (Europe/South Pacific); Herbert Sorkin (production); Paul R. Stewart (marketing)

HOME REPAIR AND IMPROVEMENT

Editorial Staff for Roofs and Siding

Editor — William Frankel
Assistant Editors — Gerry Schremp, Mark M. Steele
Designer — Anne Masters
Picture Editor — Adrian G. Allen
Associate Designer — Kenneth E. Hancock
Text Editors — William H. Forbis, Lee Greene, Lee Hassig, Brian McGinn
Staff Writers — Thierry Bright-Sagnier, Stephen Brown, Steven J. Forbis, Lydia Preston, Brooke C. Stoddard, David Thiemann
Art Associates — George Bell, Mary Louise Mooney, Dale Pollekoff, Richard Whiting
Editorial Assistant — Eleanor G. Kask

Editorial Production

Production Editor — Douglas B. Graham
Operations Manager — Gennaro C. Esposito, Gordon E. Buck (assistant)
Assistant Production Editor — Feliciano Madrid
Quality Control — Robert L. Young (director), James J. Cox (assistant), Daniel J. McSweeney, Michael G. Wight (associates)
Art Coordinator — Anne B. Landry
Copy Staff — Susan B. Galloway (chief), Lynn D. Green, Celia Beattie
Picture Department — Rose-Mary Hall-Cason
Traffic — Jeanne Potter

Correspondents: Elisabeth Kraemer (Bonn); Margot Hapgood, Dorothy Bacon, Lesley Coleman (London); Susan Jonas, Lucy T. Voulgaris (New York); Maria Vincenza Aloisi, Josephine du Brusle (Paris); Ann Natanson (Rome). Valuable assistance was also provided by: Judy Aspinall, Karin B. Pearce (London); Carolyn T. Chubet, Miriam Hsia, Christina Lieberman (New York); Mimi Murphy (Rome).

THE CONSULTANTS: Frank C. Gorham, the general consultant for this book, worked for 18 years as chief estimator for a large Virginia roofing company. A former executive secretary of the Virginia Association of Roofing and Sheet Metal Contractors, he is now a roofing consultant for architects and builders.

Claxton Walker, a home builder and remodeler for 18 years in Maryland, Virginia and the District of Columbia, now inspects homes for prospective buyers. A former industrial arts teacher, he lectures widely to real estate brokers and the public on spotting troubles in older homes. He has written magazine and newspaper articles and co-authored books on home inspection and residential energy conservation.

Harris Mitchell, special consultant for Canada, has worked in the field of home repair and improvement for more than two decades. He is Homes editor of *Today* Magazine and author of a syndicated newspaper column, "You Wanted to Know," as well as a number of books on home improvement.

Roswell W. Ard, a civil engineer, is a consulting structural engineer and a professional home inspector. He has written professional papers on wood-frame construction techniques.

Richard Ridley, an architect, founded an architectural firm that has won awards for urban planning and residential building.

R. Daniel Nicholson Jr. is an assistant production manager and estimator for a Washington, D.C., home remodeling service.

For information about any Time-Life book, please write:
Reader Information
Time-Life Books
541 North Fairbanks Court
Chicago, Illinois 60611

©1979 Time-Life Books Inc. All rights reserved.
No part of this book may be reproduced in any form or by any electronic or mechanical means, including information storage and retrieval devices or systems, without prior written permission from the publisher, except that brief passages may be quoted for reviews.
Third printing. Revised 1981.
Published simultaneously in Canada.
Library of Congress catalogue card number 77-090094
School and library distribution by Silver Burdett Company, Morristown, New Jersey.

TIME-LIFE is a trademark of Time Incorporated U.S.A.

Contents

The Complex Husk of a House

A roof repair. Valley flashing *(pages 16-17),* the metal strip under the shingles along the line where two roof slopes meet, channels floods of water every year and is prone to leaks—but replacing it is easy. Working from an extension ladder *(foreground)* hooked to the roof, the repairman slides out the old flashing and slides in a new strip, cut with tin snips from a roll of flashing metal *(left)* and shaped to fit the valley.

The greatest enemy of roofs and walls is water. It has its merits in brooks and bathtubs and drinking glasses, but in the form of rain and snow and ice it can be demoniacal. Pelting water penetrates and infiltrates. Water driven by wind or icing or capillary action defies gravity and flows upward, doing damage in hidden places.

Through the ages men have searched for ways to build waterproof walls and roofs. The ideal surfacing material would be as seamless as glass, but only masonry—brick, stone or stucco—has ever approached this ideal, and then only for siding. In applying every other material, men have followed the admirable pattern of the husk of an ear of corn, which protects the kernels from rain and sun with overlapping layers of leaves.

An ordinary shingle—whether on a roof or a wall, whether of asbestos, asphalt, wood, slate or tile—illustrates the principle of the husk on the surface of a house. Water cannot flow up the back of a shingle that overlaps the one beneath it. Lateral overlap, in staggered joints between shingles, prevents leakage from sideways flow. A soundly shingled surface tames water, causing it to meander in rivulets down the overlap of one shingle, dropping to the overlap of the next, until it reaches the bottom course and falls to the ground. The principle of overlap explains why you apply siding and roofing from bottom to top, and not the other way around.

Overlapping is also the key to roll roofing and to clapboard, aluminum and vinyl siding. The materials are laid on in long, overlapped horizontal strips; they need only a few vertical joints, which can be staggered. In aluminum and vinyl siding and in roll roofing, these vertical joints are also overlapped. At the ridge of a roof there is a double overlap, in which pieces of material that overlap each other are also laid to overlap the top course on both sides of the roof. The principle of overlap is all-important in the installation of flashing *(pages 16-21),* the metal strips that waterproof the joints of a roof. Where a roof meets a vertical surface, such as a dormer or a chimney, the flashing actually overlaps itself as well as the roof: counterflashing, fastened to the vertical surface, overlaps a channel-like base flashing, fastened to the roof; working together, the two layers of flashing direct the water safely down toward the eave.

Overlapped flashing, roofing and siding make up the surface of a house's husk. For some jobs, you may have to work with the deeper layers that lie below the surface. Shingles and clapboard are supported from below by fiber or plywood sheathing, which in turn is nailed to studs or rafters. In addition, many walls and roofs have a special barrier to dampness: asphalted building paper, nailed to the sheathing—and overlapped, of course.

The Surfaces and Structure of Roofs and Walls

For everyday purposes, the exterior of a house can be adequately described by naming the materials that cover the walls and roof—a wood clapboard wall, an asphalt shingle roof. But anyone who undertakes an extensive remodeling or repair needs a larger vocabulary. The skin of a house is a web of interlocking parts, broken at dozens of points by windows, doors, vents and roof projections, and supported by a hidden framework of rafters, joists, studs and beams *(pages 10-11)*. Each part has its precise name, and to discuss your plans with an architect or building inspector or to order materials from a supplier, you should learn some of the language of the trade.

Though roofs are made in dozens of styles and covered by a variety of materials *(page 67)*, most roof shapes are variations on two types: the gable roof and the hip roof *(right)*. The gable roof, sloping on two sides, is named for the triangular wall section, or gable, formed by the slopes at each end wall. The horizontal upper edge of the roof, where the two slopes meet, is the ridge; the horizontal lower edges are eaves; and the sloping edges that frame the gable are rakes.

A hip roof has slopes on all four sides and eaves all the way around; the roof is named for the raised corners, or hips, at which slopes intersect. The fourth slope on the hip roof at right has been eliminated to illustrate a valley—the V-shaped channel or trough formed at the intersection of two slopes.

Different roof styles are, of course, often combined in a single building. A gable-roofed home may have an extension covered by a flat roof *(page 84)* or an enclosed patio with a single-slope shed roof *(page 106)*. But these differing styles remain variations of the basic ones: a flat roof is one with no slope at all; a shed, or lean-to, roof is actually half of a gable roof; a gambrel roof *(not shown)* is a gabled roof with a steep lower slope and a flatter upper slope on each side of the ridge; and a mansard roof, which has two slopes on each of its four sides, is essentially a gambrel roof with hips.

Roof slopes are broken not only by intersections with other slopes but by projections that admit light and air to an attic or release smoke, fumes and hot air from the house. Dormers—roofed projections designed to house vertical windows—are described by their roof style: gable, hip or shed. Ventilation openings are named for their locations on a roof: ridge vents provide a continuous opening along the ridge line, gable vents are mounted high in an end wall and roof vents are set into the slope of the roof. Chimneys may be set into the roof slope or built into a gable wall; a stack vent—a 3- or 4-inch pipe that releases gases from the plumbing system—usually rises from one of the main slopes near the ridge.

Any substantial interruption in a roof—the joints around projections and the seams where roof slopes meet or where a roof meets a wall—must be waterproofed with metal flashing. Flashing comes in a variety of forms, including a collar that fits around vent pipes, a wide strip that runs along the full length of a valley and a double-layer pattern, called counter- and base flashing, used at seams where roof slopes meet walls or chimneys.

At their eaves, most roofs project well beyond the building wall to carry rain away from sidings and foundations. The projection, called the overhang or cornice, usually consists of a fascia board nailed to the rafter ends and a soffit board that forms the underside of the overhang. Many gable roofs also overhang at the rakes: the underside, like that of an eave, is a soffit; the outer face is called a rake board or rake fascia. Rain gutters usually hang from the eaves fascia, and metal drip edges along eaves and rakes direct runoff away from the fascia boards. In colder climates, snow guards mounted above the eaves prevent frozen snow from sliding off the roof in sheets.

Because it is protected by overhangs and gutters from the direct attack of sun, wind, rain and snow, siding is less trouble-prone—and less complex—than roofing. The chart on page 25 lists a variety of available siding materials; the drawing at right shows one of the most common: overlapping horizontal boards, called clapboard or lap siding.

The structural terms and features of siding are also few and relatively simple. The gap between the siding and the soffit is finished with a frieze board, and the lower edge of the siding extends below the top of the foundation wall. Siding edges are covered by corner boards at outside corners and butted against corner strips or moldings at inside corners. Metal drip caps over door- and window frames protect the casings from water damage, and door- and window sills are sloped outward for drainage.

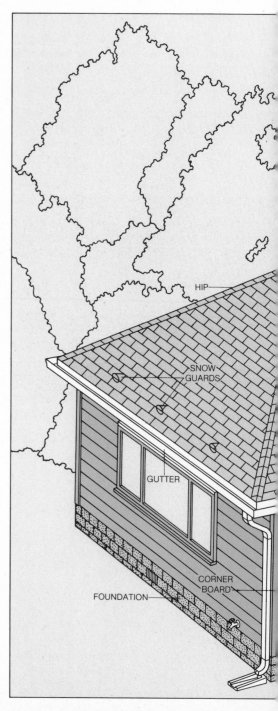

HIP

SNOW GUARDS

GUTTER

CORNER BOARD

FOUNDATION

An unlikely exterior. This improbable house is a composite of clashing styles and features; the purpose of the drawing is not to represent a real house, but to illustrate the exterior parts defined and described in the text at left. Every detail, however, is accurately rendered, and mastering the names of these parts is an essential first step in any job of repair or remodeling.

BASE FLASHING
COUNTER FLASHING
ROOF VENT
STACK VENT
RIDGE
OVERHANG
SKYLIGHT
RAKE
FRIEZE
GABLE VENT
RIDGE VENT
GABLE
VALLEY
SOFFIT
VALLEY FLASHING
DRIP EDGE
RAKE BOARD
GABLE DORMER
FASCIA
EAVE
DOOR CASING
DRIP CAP
WINDOW CASING
INSIDE CORNER
SILL
SIDING

The Skeleton under the Skin

Before you can cover a house with roofing or siding, you must know the skeleton that lies beneath the exterior—what builders call the framing. The type of framing depends on whether you have a masonry house or a wood-frame house.

A wood frame (*opposite*) consists entirely of lengths of wood, named for their location and function. Normally, vertical studs frame the walls, horizontal joists support the ceilings and floors and diagonal rafters hold up the roof. These framing members are generally installed with a narrow edge facing outward or upward, and this edge provides a nailing surface for the layers of the house's skin.

Generally, the innermost layer is a sheath of plywood or fiberboard panels. A felt covering is stapled to the sheath as a barrier against moisture and air, and the siding (*pages 24-25*) is nailed through the sheath and into the studs.

The spacing of the studs determines the nailing pattern of sheathing and siding. In most frame houses studs are either 16 or 24 inches apart, but the spacing may vary between a full stud and the shorter cripple and jack studs that are used to frame doors and windows, or between a stud and a corner post.

The spacing problem takes a different form in a masonry house (*below*). Here, the skeleton consists of solid brick, stone or concrete blocks covered with brick veneer, stucco or ordinary siding. If siding is used, it hangs on vertical wood supports called furring strips, spaced 16 or 24 inches apart on the outside walls.

In most modern dwellings the roofs are built with joists and rafters (*below and opposite*) and decked, or sheathed, with boards or plywood panels. Depending on the roofing material, boards may be butted together, tongue and grooved or spaced up to 4 inches apart. The boards or panels are nailed to the rafters and covered with felt paper; flashing is then installed at chimneys, vents, dormers, skylights and valleys (*pages 16-21*).

The final roof covering (*pages 66-67*) is as variable as the siding on the walls. On a pitched roof, the covering may be asphalt shingles, tiles, slates, wood shakes or shingles; a roof that is flat or has negligible pitch must be covered with metal panels, built-up or roll roofing.

A masonry house. In this example of masonry construction, a hip roof covered with tiles rests on concrete-block walls finished with stucco. To anchor the roof framing to the walls, a wood sill plate is bolted to the top row, or course, of concrete blocks. Hip rafters extend from the sill plate to the horizontal ridge board; these rafters are supported by hip jack rafters fastened to the sill. The main part of the roof is also supported by common rafters, which are fastened to the ridge board and the sill plate. Ceiling joists span the width of the house, resting on the sill plate and nailed to either jack or common rafters. Horizontal braces called collar beams tie opposite rafters to give the roof added strength.

The roof sheathing is 4-by-8-foot plywood sheets nailed to the rafters and covered with roofing felt and tiles. On the walls, two coats of mortar create the stucco exterior: a thick base coat, applied directly to the blocks; and a finish coat, spread over the base. (Stucco can also be applied to a wood-frame house in three coats, by the method described on pages 54-61.)

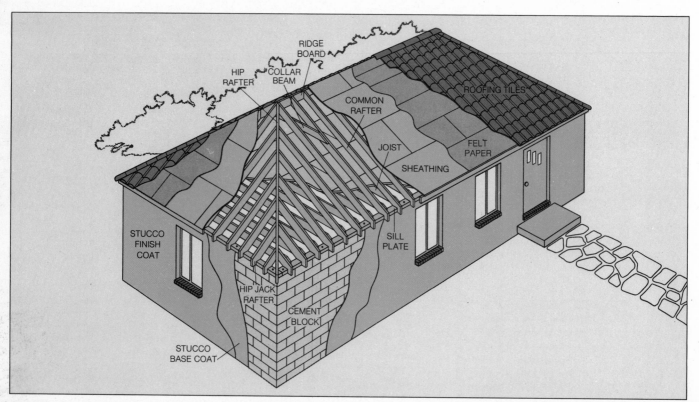

A wood-frame house. Beneath the clapboard siding and asphalt shingles of this typical frame house, vertical wall studs and sloping rafters make up the main structural members. The studs rest on a horizontal 2-by-4 called a sole plate, and a horizontal top plate consisting of doubled 2-by-4s runs across the top of the wall studs. The lower ends of the rafters rest on the top plate; the upper ends are attached to a horizontal ridge board. Horizontal ceiling joists, nailed to the rafters, rest on the top plate, and collar beams add rigidity to the structure.

Special framing is needed wherever walls are interrupted by windows or doors and wherever roof slopes or walls intersect. In the roof, the valley is framed with a valley rafter, which runs from the ridge boards to the top plate and is supported on each side by shorter valley jack rafters. In the walls, windows and doors are framed horizontally with headers and plates, vertically with jack and cripple studs. Corner posts consist of three studs nailed together, and each gable wall is framed with end studs installed between the top plate and the end rafters.

To create a solid surface for the final coverings, both roofs and walls are sheathed with plywood sheets; felt paper shields the sheathing against moisture. The valley and other critical joints are protected with metal flashing and the roofing and siding are nailed over the sheathing.

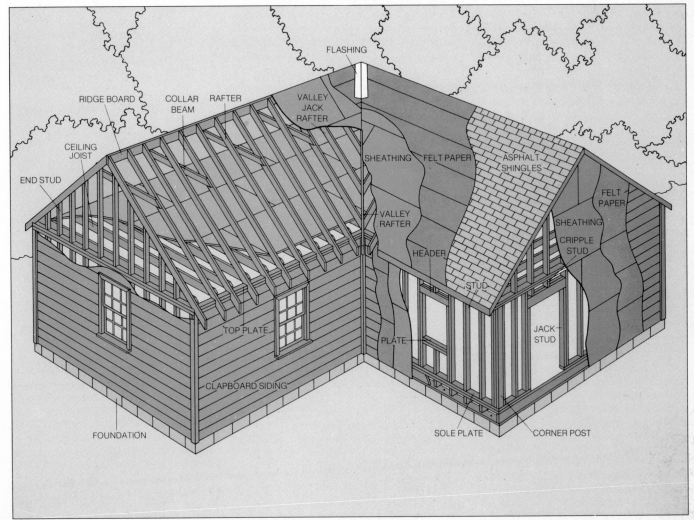

Professional Tools for Reaching High Places

Professionals have a catalogue full of devices that make climbing to, working at and lifting materials to heights of 20 and 30 feet safe and easy. Most of these tools are available at rental agencies.

The basic tool, of course, is the extension ladder. For the jobs in this book use a Type 1 ladder, the strongest commercial ladder. New ladders are marked conspicuously with this classification. If you have doubts about the strength of a ladder you own, it is wise to rent a ladder that you can count on to support you and a bundle of shingles or siding.

Though a ladder is often used as a work platform for small siding repairs, larger jobs are more efficiently completed with a scaffold that extends at least 8 feet horizontally along the wall. It saves the time spent moving ladders. You can make a scaffold from a pair of extension ladders and special brackets called ladder jacks *(below)*. A more versatile scaffold *(below, right)*, designed to be moved up and down by foot power, is a better choice because it enables you to readjust the platform position easily to a comfort-

able working height. Both these devices support working platforms of 12-foot 2-by-10 fir planks of a type sold as scaffold grade. Use them in pairs supported every 9 feet; such a platform holds two men plus about 75 pounds of materials.

Elaborate metal scaffolds, favored by contractors who may have dozens of workers scrambling over a building at the same time, are generally more trouble and expense than they are worth to a homeowner. But for those who distrust scaffolds supported by wood, as for some elaborate jobs such as stuccoing a wall *(pages 54-61)*, metal scaffolding may well be worth the cost of renting it.

In addition to a ladder and possibly scaffolding, the equipment needed for roofing includes hooks to hold a ladder to the ridge or brackets to support a foot rest *(page 14)*. Of course, if your roof slopes less than 4 inches in 12, you can work on it with no more assistance than rubber-soled shoes. For both roofing and siding work, you need support and access for materials as well as yourself. To lift the materials, says one contractor, the

hardest way is often the easiest: lug the material, a batch at a time, up a ladder. However, you can ease some of this effort by using a ladder as a ramp, renting a crank-operated elevator *(page 15)*—or enlisting a sturdy helper.

Safety Tips for Ladders and Scaffolds

Strong as the equipment on these and the following pages is, it requires careful use:

☐ Always lean a ladder against a wall so that the bottom is one fourth the distance from the wall that the ladder extends up it.
☐ Rest a ladder on level ground, or use an adjustable foot on one of the rails.
☐ Allow only one person at a time to climb a ladder.
☐ Keep metal ladders away from overhead electric wires.
☐ Have someone steady the ladder as you climb it.
☐ Stay off ladders in high winds.

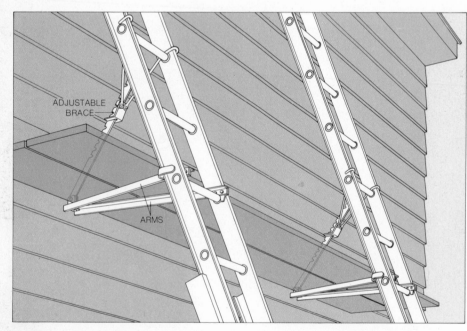

A scaffold hung from ladders. Ladder jacks, special plank supports designed to be hung from the rails and rungs of extension ladders, can support a work platform of 2-by-10s up to 9 feet long; they are useful for heights up to 20 feet above the ground. Suspend the jacks from two extension ladders leaned against the house so that the rungs are about 2½ feet from the wall at the height you wish to stand. Level the bracket arms with the adjustable brace on each ladder jack. Then enlist a helper to lay the planks so that they extend 1 foot beyond the arms.

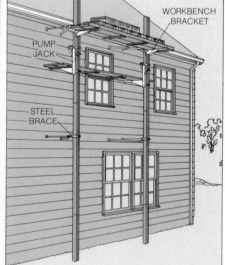

An adjustable scaffold. Movable steel brackets called pump jacks travel up and down wood uprights to position a platform of planks—accessible by ladder—at any convenient working height up to 30 feet. The uprights are usually set into shallow holes 9 feet apart anchored to the wall with steel braces *(right)*. The uprights also support accessory brackets for a guardrail or, as shown above, a workbench for materials.

Pump Jacks and Scaffold: A Portable Freight Elevator

1 **Constructing the uprights.** Fasten 2-by-4s side by side with tenpenny nails to make two 4-by-4 poles. Drive the nails into both sides of the pole, spacing them a foot apart. To make a long upright—up to 30 feet—splice boards together, staggering joints at least 4 feet.

2 **Bracing the uprights.** Braces clamped near the tops of the uprights and nailed to the wall or roof hold the uprights vertical. Clamp the braces before erecting the uprights so that the bolts cross the pole seams, squeezing the halves together; then nail the braces to the uprights. If the uprights are to stand on a hard surface, install additional braces to secure the bottom of each upright; in soft ground, dig holes for the bottoms 4 inches deep and 32 inches from the wall.

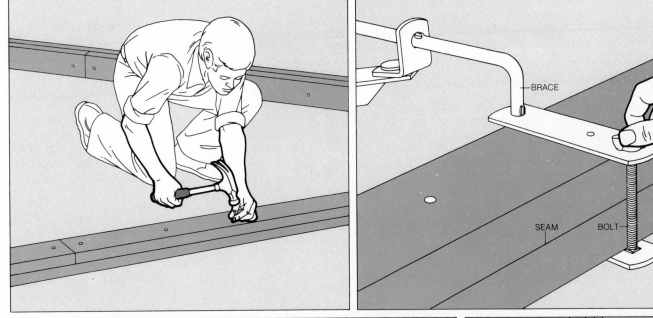

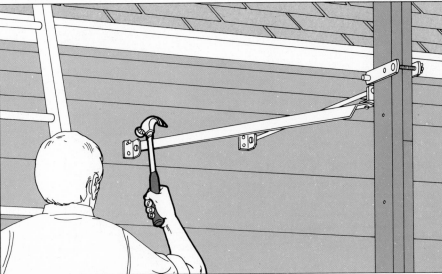

3 **Anchoring a brace to the wall.** Tilt each upright into position, and while a helper holds it, use 16-penny nails to fasten the arms of each bracket through the wall or roof into studs or rafters. (For brick or block construction, use 1½-inch masonry nails.) Anchor the longer arm of the brace between the poles (*above*), then install additional braces as necessary to support the uprights every 7 to 10 feet of their height.

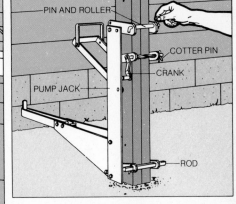

4 **Mounting the pump jacks.** Attach a pump jack and a worktable or guardrail support to each pole. To do so, pull the pin and roller, crank and stationary rod from each jack by first removing the safety cotter pins that prevent these components from working loose (*above*). Rest the pump jack on the ground and, holding it against the pole, replace the pin and roller, crank and stationary rod; secure them with the cotter pins. Worktable and guardrail supports are fastened to the poles above the pump jacks in a similar fashion but with fewer pins. Lay scaffold grade 2-by-10 planks across the jacks and worktable supports. Use a 2-by-4 as a guardrail.

5 **Operating the scaffold.** Raising or lowering the scaffold is best accomplished by two people working in unison, but you can do it alone. To lift the platform, raise and lower the foot lever on one pump jack until one end of the platform rises about a foot above the other, then repeat at the other end. When you have pumped the scaffold to the height you wish, leave the foot levers in the up position. To lower the scaffold, depress the stationary rod with your foot and turn the crank (*inset*). To jack past a brace, temporarily detach the brace from the upright and let it hang against the wall. Then raise or lower the scaffold past the brace and reattach it.

Secure Toeholds on a Sloping Roof

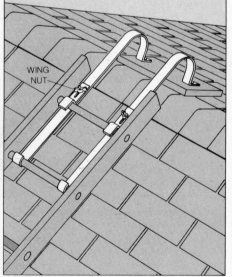

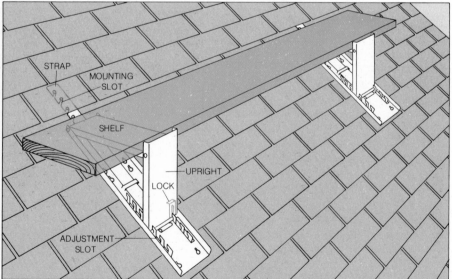

Using rungs as toeholds. Fitted with a pair of ladder hooks, a section of an ordinary extension ladder becomes a toehold for working on roofs. Simply clamp the hooks to the top two rungs of the ladder by tightening a wing nut, then suspend the ladder from the ridge as shown above. A wood block under the hooks spreads weight to prevent damage to shingles.

A roof-top scaffold. A pair of triangular metal platforms called roof brackets provides level storage or working area on asphalt shingles (and uncovered roof sheathing). Each bracket consists of a steel strap with slots or holes in one end for fastening to the roof. Attached to the strap is a shelf, braced by an upright, which supports a 2-by-10 plank. Adjustable brackets like the

ones shown above can fit any roof pitch and have a lock to keep the upright from slipping. To attach a bracket, bend back a shingle tab, insert an eightpenny nail in a bracket slot (or hole) and drive it through the shingle below. Remove a bracket with a hammer by knocking it toward the ridge and slipping it off the nail, then lift the shingle tab and pound the nail flush.

Two Ways to Hoist Heavy Loads

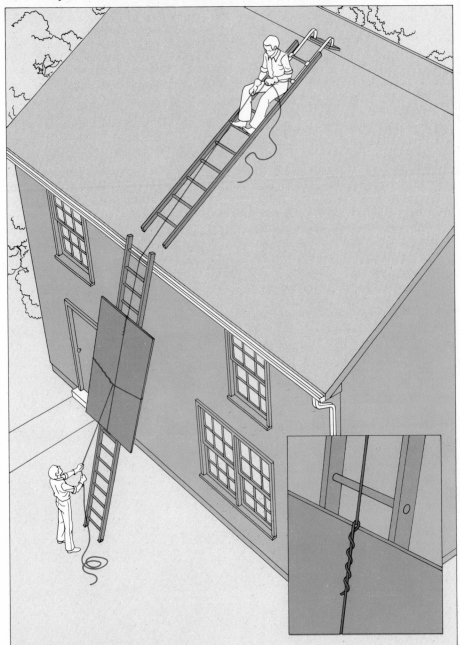

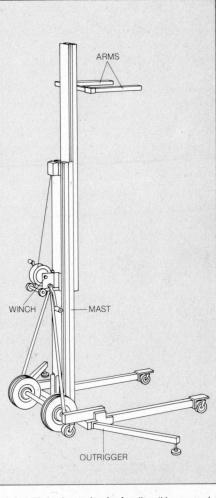

ARMS

WINCH — MAST

OUTRIGGER

Hauling materials up a ladder. An extension ladder can serve as a ramp to steady a heavy or bulky object as you pull it up to a scaffold or a roof with a rope. Tie a length of half-inch rope around the load with a timber hitch *(inset)*. Climb onto the scaffold or roof and pull the load up the ladder. For material as large as the 4-by-8-foot sheet of plywood sheathing *(above)*, add a guide rope to be held by a helper on the ground.

A forklift for heavy loads. A collapsible, manual forklift from a tool rental agency can lift a 500-pound load of materials as high as 26 feet to a scaffold or an eave. The fork has adjustable arms to carry almost any size load and is cranked up the mast with a self-locking winch. (Small objects can be stacked on a plywood platform lashed to the arms.) The mast on this model automatically telescopes upward as the load is raised; on other types of forklift, the mast must be preassembled to the desired height. When using a forklift, be certain to set it on level ground and to extend the outriggers, if there are any, to prevent the forklift from tipping over. Also see that the casters are chocked to keep the forklift from rolling.

Waterproofing the Weak Spots with Metal

Because roofs are most vulnerable to leaks where their slopes are broken by chimneys, dormers, vent pipes and valleys—the joints where sections meet at an angle—these places should be covered by a continuous watertight material called flashing. Roll roofing *(page 82)*, galvanized steel, tin and terneplate—sheet metal coated with a tin-lead alloy—have been widely used for flashing. The best materials are aluminum and copper; both are durable, corrosion-resistant and easily worked. Of the two, aluminum is generally less expensive and more available for home use than virtually indestructible copper, often favored for public and commercial buildings.

Old flashing should be removed and replaced when it begins to leak or show signs of deterioration. The job is time-consuming but not difficult, requiring ordinary hand tools and sturdy ladders. Be prepared to remove and replace shingles that often lap—and sometimes completely cover *(page 81)*—existing flashing. It is also good practice to install new flashing directly on top of old material when reroofing *(page 78)*. In that way, you avoid the risk that the old flashing will spring leaks between roofing jobs.

Ordinary flat flashing is available from roofing suppliers in rolls (usually containing 50 feet) and shorter strips in various widths. The narrow widths, 6 to 8 inches, are generally used as base flashing along the sides and lower edges of chimneys; wider sheets, 14 to 24 inches, are used for valleys and at the up-roof edges of chimneys *(page 19)*.

Valley flashing should extend over the ridge cap at the upper end and slightly below the eave at the lower end, so add at least a foot to valley measurements to allow for these overruns, plus an extra 4 inches for overlapping if more than one length of flashing material is needed.

Flashing is so thin—about 1/50 inch—that it is easily cut with metal shears and may be bent by hand into the trough shape required for valleys by a technique like the one used to bend aluminum siding *(page 45)*. However, some special bends—used for extra watertightness in valleys—are best made to your order by a sheet-metal shop. One commonly used valley type has its long edges bent inward; called slater's edges, they form a water barrier on either slope of the valley and provide secure grips for the cleats that hold the flashing in place. Another type, the W valley, is used when the pitches of adjoining roof sections differ radically. It has a ridge 1½ to 2 inches high along the center of the valley to prevent water from the steeper slope from running up the lesser slope.

New Flashing for a Valley

1 **Removing the ridge cap.** Starting from the top of the valley, using a pry bar to remove asphalt or wood ridge shingles covering the end of the flashing *(above)*. For tile roofs, use a cold chisel and ball-peen hammer to chip away the mortar under the cap tiles *(inset)*. On slate roofs, remove shingles that meet at the ridge. Remove nails or cleats that hold flashing to the ridge.

2 Clearing away the shingles. Free the edges of the old flashing by loosening or removing overlying shingles. The ends of asphalt shingles can usually be lifted clear by removing nails closest to the valley *(right)*. Use an old kitchen knife or a long-bladed trowel to cut away adhesive.

Brittle shingles—slate, tile, asbestos, wood and even cold asphalt—may require the use of a slate puller *(inset)*, available at roofing suppliers, or a hacksaw blade *(page 48)*, to cut concealed nails. Slip the arrowhead-shaped tip of the puller under the overlying shingle until it hooks a nail. Then hammer downward on the handle to shear the nail.

Work downward as you clear away shingles, prying out nails or cleats along the flashing edge. Pull down the old flashing from the bottom.

3 Sliding the new flashing. Slide new flashing upward under the loosened shingles. Where roof pitches are uneven, use W-valley flashing with slater's edges and a crimped center barrier *(inset)*. If you use two sections, position the lower one first with a pair of nails an inch from the top. Make sure that the top section overlaps by 4 inches, with its slater's edges inside the ones below, before nailing to the ridge.

SLATE PULLER

CLEAT

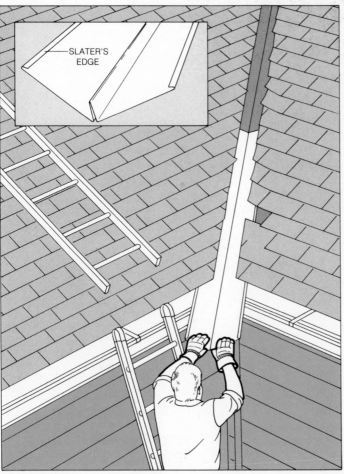

SLATER'S EDGE

4 Fastening the cleats. If the flashing has slater's edges, hook pairs of cleats over the edges every 4 feet and nail them an inch from the valley edges. Fold cleat ends to cover nailheads. With plain-edged flashing, nail at the ridge and seal edges with roofing cement.

Restoring the Roofing

Renailing asphalt shingles. Starting from the bottom, press and nail shingles flat along the valley edges. Do not use old nail holes. Find new spots near them and use asphalt cement to fill the old holes and the spaces between shingles. Do not nail or cement shingles to flashing.

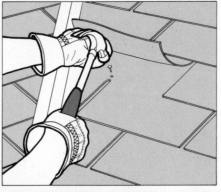

Replacing wood shingles. Use a hammer and wood block to tap the replacement into position. Then nail it to the roof as close as possible to the edge of the overlying course and cement the nailheads.

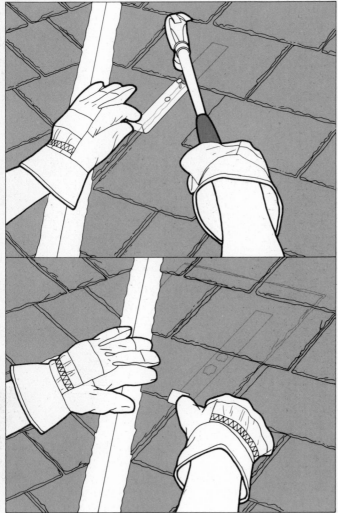

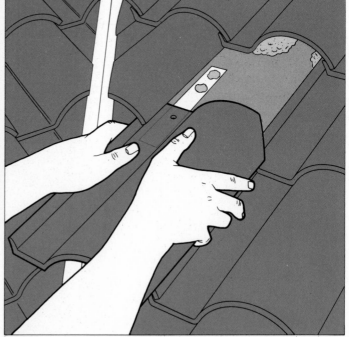

Replacing slates. Since brittle materials cannot be bent back for nailing, make a special holding tab by cutting a strip of metal flashing material 2 inches wide and long enough to extend about an inch below the replacement when inserted under the course above. Nail the tab into the roof through the joint of the underlying course (*top*) and cover the nailheads with roofing cement, then slip the replacement into position over the tab. Bend the projecting end of the tab upward and press it down firmly over the bottom edge of the slate (*bottom*).

Replacing tiles. Mount a holding tab, like that used for replacing slate, onto the roof. Then add two dabs of roofing cement directly below the overlying course before sliding the replacement tile into position (*above*).

Two Layers of Flashing for Flexible Joints

A more elaborate type of flashing, consisting of two layers of material, is usually required where a roof meets a vertical surface—chimneys, dormers, walls—and is most leak-prone. The first layer, called base flashing, is fastened to the roof and extends along the vertical surface; it is partly covered by a layer of counter flashing *(page 20)*, fastened to the vertical surface. This arrangement provides a flexible joint, allowing the roof to shift slightly with expansion and contraction.

2 Interlayering side flashing. At the down-roof end of the chimney, set the first piece of base flashing under a shingle so that it extends beyond the down-roof edge. Lap each successive piece over the one below, interlayer it with shingles, then nail it to the roof. The last piece should extend beyond the up-roof edge.

3 Flashing up-roof. Cut a sheet of flashing material 18 to 24 inches wide and a foot longer than the chimney width. Bend it lengthwise about 4 inches from one side, form a pair of 6-inch tabs *(right)* and fit it with the narrow side against the chimney and the wide side forming an apron against the roof. Push the apron under the shingles and drive nails in the corners where they will be covered by shingles *(inset)*. Fold the tabs over the base flashing at either end, then seal all joints with roofing cement.

Sealing a Chimney

1 Base flashing for the sides. Clamp pieces of flashing, each 7 inches long and about 6 to 8 inches wide, between wood blocks. Using a mallet, bend each piece so that half will lie against the chimney, half against the roof. Use the same bending method for other base flashing.

APRON

TAB

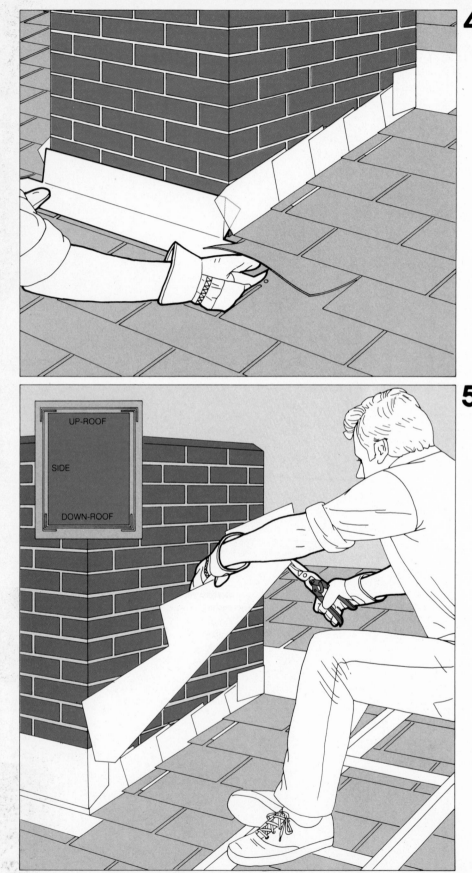

4 Flashing down-roof. Prepare down-roof flashing like that used up-roof, making it the same length, bending it down the center and folding the tabs inward, but using material only 6 to 8 inches wide. Position it against the chimney, slipping the tabs behind side flashing and the projecting ends of the apron under shingles. Fold the ends of the side flashing over the down-roof flashing. Nail and seal as for up-roof flashing.

5 Counterflashing a chimney. Cut four pieces of 8-inch flashing for the four sides of the chimney, making each long enough to wrap about 6 inches around the corners. Cut steps in the pieces for the slanting sides so that, when the bottoms of these pieces lie about ¼ inch above the roof, the tops line up with the brick joints. The up-roof and down-roof pieces need no steps. Beginning down-roof, nail each piece into the joints with masonry nails, always lapping the one below (inset). Seal the upper edges of the counterflashing.

Special Techniques for Dormers, Walls and Pipes

Joining two valleys. When two valleys intersect at a ridge, as at the back of a dormer, fit the valley ends of the flashing snugly over the ridge and each other by making small cuts and bends along the edges *(inset)*. Cover the overlapping ends with shingles.

Flashing the base of a wall. Where a roof meets a vertical wall, as at the sides of dormers, base flashing is installed in the same way as with a chimney *(page 19)*. But counterflashing techniques may vary with the wall material. If the wall has lapped siding *(left)*, slip pieces of metal counterflashing upward under laps and nail them in place. Stucco walls, which contain corrosive alkalies, should be coated with asphalt cement before counterflashing is nailed to the surface. Smooth wood walls can be notched with a saw to hold the top edge of the counterflashing. Shingled walls require no counterflashing; the shingles overlap the base flashing to provide adequate protection.

Sealing vent openings. Specially designed pipe flashing is used to protect vent openings. After removing the shingles and the old flashing, and sealing any nail holes with roofing cement, simply slide the new flashing over the pipe until it is flush with the roof. Nail it down and replace the shingles. If a protective collar is not provided for the joint between the pipe and flashing, caulk the joint thoroughly.

BASE FLASHING

COUNTER FLASHING

2

New Coverings for Old Walls

Two ways to set the courses. The aluminum siding at left interlocks in horizontal courses up a wall; in effect, the manufacturer determines the height of a course. Calculating courses for shingles or clapboard is trickier. The installer should make a story pole *(page 49)* from a long 1-by-2, marked for even courses from the bottom of the wall to the top. To mark the pole, he needs the compass-like carpenter's tool called dividers, and to transfer the marks to the wall, he needs a chalked string. The windup reel shown here contains its own powdered-chalk supply and a 50- or 100-foot string.

The siding of a house should be weathertight and durable, and these qualities are surprisingly easy to attain; unlike the roofing or plumbing, good siding should last the life of the house. But siding is more than protection. Along with architectural style, it is the major factor in the look of a house and changing it can dramatically transform the building's appearance. You may from time to time need new siding to replace damaged or deteriorated material, but you may also want it simply to make your house look better.

Putting the siding up is mostly straightforward, uncomplicated work. Remember, though, that you cannot match a professional's speed and dispatch, especially if you can work only after hours and weekends. Be prepared, therefore, for a spell of disruption in your life. For a time, you may have to do without downspouts, outside electrical fixtures and, briefly, power, gas or air conditioning.

Because you will not be able to apply your new siding fresh off the delivery truck, you must take pains to store it well. Durable as siding is on a wall, it is highly vulnerable on the ground. A hard rain will soak through the paper bags of the cement and lime used to make stucco, hardening the cement and spoiling the lime; wet sand will add an unmeasurable amount of water to the standard recipe for a stucco mix, depriving it of strength. Any wood—plywood, hardboard, clapboard, shingles—will warp if the water content of one side exceeds that of the other; not even the weather resistance of cedar exempts it from this law of nature. Rain also spatters mud on materials and encourages mildew, in both cases creating blotches that are troublesome to remove.

Move your automobile out of the garage and store all these vulnerable materials there. If you do not have roofed storage space, make outdoor platforms out of rows of concrete blocks that support 2-by-4s, spaced widely enough to let air circulate freely under the stacked materials, and cover the stacks with heavy-duty polyethylene sheeting tacked or weighted so that it will not blow away.

The elements will not normally damage aluminum and vinyl sidings, but thoughtless handling can. Aluminum is easily dented or scratched. Vinyl is tougher, but in freezing weather it becomes brittle and may snap or crack if sharply struck. Leave both of these materials in their protective cartons until just before you use them, and store the cartons in an out-of-the-way place, under cover—rain-soaked cardboard will break and spill the contents, to their detriment. Asbestos shingles are brittle in any weather; if you drop a bundle of them, every shingle may break. Keep them away from household traffic and, as always, under cover; they stain if left wet for more than a day or two before they go on the wall.

Choosing the Right Siding, Preparing the Walls

Beyond the obvious considerations of cost and appearance, your choice of a new siding should take into account the problems of installing a specific siding material, the kind of regular maintenance it will need, and its special physical characteristics, which may make it more or less suitable for your purposes.

The drawings and chart on these pages enable you to compare eight widely available materials. In the drawings, you can compare their surface appearances and judge their suitability to the architecture of your home.

The appearances may be deceiving. Clapboard, stucco and wood shingles and shakes are traditional materials that have set the familiar styles in exterior wall coverings for centuries. The other sidings, developed more recently, are usually shaped and finished to resemble the traditional ones. At their best, these imitations are not wholly successful: the newer types of siding lack the elegance of the old materials. But what they lack in appearance they may make up in economy or convenience. Vinyl, for example, has become a favored material for new siding on both new and old houses because it is relatively inexpensive, easy to install and all but maintenance free. It is factors like these that are compared in the chart at right. Detailed descriptions of all the sidings appear in the remainder of this chapter.

One element of a siding job that does not show up in the drawings or the chart is the exterior wall beneath the siding. Depending on its condition and on the siding you choose, this wall may need some measure of repair or preparation, by methods described on pages 26-32.

Eight Styles of Siding

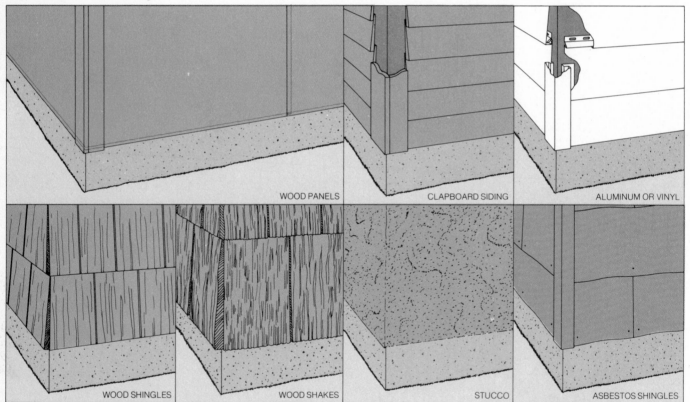

WOOD PANELS

CLAPBOARD SIDING

ALUMINUM OR VINYL

WOOD SHINGLES

WOOD SHAKES

STUCCO

ASBESTOS SHINGLES

A range of materials. The types of siding shown above are arranged in the order in which they are discussed in this book. Some look alike, but each has distinctive qualities. Large, rectangular panels of plywood or hardboard are available unfinished for painting or staining, or in finishes that resemble any of the other materials. Long planks nailed horizontally to walls can be installed in a variety of patterns. Clapboard, the most common, consists of lapped boards tapered toward one edge; in tongue-and-groove sid-

ing the board edges are fitted together, and in shiplap they are rabbeted and overlapped. In each case the boards can be painted. Vinyl and aluminum are generally designed to resemble clapboard when installed, but come from the factory precolored in several shades.

Wood shingles and shakes are similar in size and identical in material—redwood or cedar—but shingles are milled to an exact and uniform size while shakes are thicker and irregularly

shaped. Both can be stained or painted, or in the case of cedar, simply left unfinished for an attractive, weathered look. Asbestos-cement shingles, usually molded to look like wood, are most often sold precolored. Masonry siding, represented here by stucco, is made from a wet cement that is spread over a wall in layers. Stucco may be left uncolored; more often, it is either precolored or painted after installation. Its surface may be smooth, lined and grouted, or embedded with light-colored stones.

A Guide to Siding Materials

Siding type	Cost	Maintenance	Advantages	Limitations
wood panels	inexpensive (un-finished plywood) to moderate (fin-ished hardboard)	regular painting or staining	quick installation; goes over most existing sidings; available in a wide variety of styles	poor fire resistance; installation always requires two workers, and can be especially difficult at the borders of windows, doors and rake
clapboard	moderate	regular painting or staining	goes over most existing sidings	poor fire resistance; installation requires two workers; some types are subject to rot
vinyl	inexpensive to moderate	none	easy installation; goes over most existing sidings	may melt near intense heat; brittle in very cold weather; narrow range of colors, subject to fading; cannot be painted
aluminum	inexpensive to moderate	none	easy installation; goes over most existing sidings; fire resistant; available in wide variety of styles; can be repainted	scratches and dents easily; may clatter in wind and hail if not insulated
wood shingles	expensive	regular replace-ment of missing or damaged pieces; regular painting or staining for some woods	goes over most existing sidings; single pieces easily replaced; can be left unfinished for rustic look	flammable; slow installation
wood shakes	expensive	regular replacement of missing or damaged pieces	goes over most existing sidings; single pieces exceptionally durable and easily replaced; can be left unfinished for rustic look	flammable; slow installation, often difficult around windows and doors
stucco	moderate	none	fire resistant; surface can be molded or decorated	requires wire-mesh backing over wall or existing siding; long and difficult installation, requiring special skills and caustic materials, must be done in good weather; cracks or crumbles if incorrectly applied
asbestos shingles	inexpensive	regular replacement of missing or damaged pieces	easy installation and replacement of pieces; fire resistant	must usually strip old siding before installing; cracks and chips easily

Comparing siding materials. In the chart above, "Cost" refers to the relative cost of each material as compared with all the others; it does not reflect the labor involved in a professional installation. In general, labor costs are considerably higher for the traditional materials—clapboard, wood shingles and shakes and stucco—than for the newer types of siding, most of which are relatively easy to install. With stucco, the materials alone are only moderately expensive, but if professionally installed, it is the most expensive of all sidings. For some materials a range of costs is given. Aluminum, for example, can be bought in two thicknesses, each with or without an insulation backing. Thin aluminum without backing is inexpensive; thick aluminum with backing is moderately expensive.

With reasonable care, all the materials listed will last as long as the houses they cover. The column headed "Maintenance" indicates what must be done to keep a siding structurally sound and weatherproof. It does not take into account the gradual deterioration of materials like aluminum and vinyl, which, though sound, may look weathered after several years.

The last two columns summarize the general advantages and limitations of each material. Under these headings the most important considerations for the amateur are ease of installation and the ability to cover existing siding.

A Solid Base for New Siding

The principal requirement for any type of siding is a flat, sound nailing surface and there are three ways of obtaining it. The old wood siding will do if it is in good condition. If not, the usual solution is to mount furring strips over the old siding—shimming the strips on uneven surfaces—and nail the siding to the strips. The third and most time-consuming way to guarantee a good nailing surface is to remove the old siding and begin as you would with new construction. This often must be done when the old siding is asbestos shingles, crumbling stucco or aluminum—all difficult to nail through to a firm wood backing.

Whichever technique you choose, always begin by removing accessories that could interfere with the re-siding—downspouts, light fixtures, shutters, decorative trim. You may need professional help if you must temporarily remove a utility meter. It may be necessary to shut down all power; in that case be prepared to mount the necessary pieces of siding quickly so power can be restored.

Preparing the walls for re-siding offers an excellent opportunity to add to wall insulation. If you plan to keep the old siding in place, you can drill holes through it to blow loose-fill insulation into the stud cavities, or you can mount plastic foam boards between old and new siding. If old siding is removed, install batts or blankets between the studs.

Removing asbestos shingles. Because brittle asbestos shingles provide a poor nailing surface for new siding, strip them from the walls with a pry bar. Break the shingles near the nailheads and pull off the pieces and backer strips; then pound the nails into the sheathing. Use 15-pound asphalt felt to patch any old felt torn in the process, making sure the new paper overlaps the old by about an inch.

Stripping the Walls

Removing trim. To save molding and other trim for reuse, pry it off gently with a utility bar. Begin at one end, inserting the flat end of the bar under the molding and tapping the curved end with a hammer. Remove the frieze only if you plan to replace it with siding. Otherwise, build out the old frieze with an identically sized strip of lumber to meet the projection of the new siding.

Resealing the Walls

Sheathing a stud wall. If studs are exposed after removing old siding, add any necessary insulation—with the vapor barrier inside—before sheathing the wall with C-D exterior plywood or fiberboard. Mount 4-by-8-foot sheets horizontally for greater rigidity, leaving about ⅛ inch between for expansion. Space nails 6 to 12 inches apart on the studs. Do not overlap the board ends at corners, but caulk the seam. Overlap the house foundation by at least an inch.

Sheath behind window casings and doorframes by inserting sheathing into space from which old siding has been removed, or by prying off the trim and renailing it over the sheathing.

Nailing sheathing paper. For additional moisture protection, especially when siding with stucco (*page 54*), or asbestos cement shingles (*page 62*), staple or nail a layer of 15-pound asphalt felt to the sheathing. Attach the paper in long horizontal rows from the bottom, lapping each row about 4 inches over the row below. Be sure to mount strips of sheathing paper, at least 8 inches wide, around windows and doors—even if none is used elsewhere—to reduce drafts.

New Insulation over the Old Siding

Attaching plastic foam boards. If the interior wall has no vapor barrier—to keep warm inside air from condensing in the wall—drill half-inch vent holes through the old siding and sheathing (*left*) at the top and bottom of each stud cavity. Nail 2-by-8-foot panels horizontally on the siding. Leave about ¹⁄₁₆-inch space at corners so that trapped vapor can escape.

Furring for Panels and Horizontal Siding

1 **Locating studs.** Find and mark the tops of studs by examining the nailing pattern in your present siding, removing pieces if necessary. If you are uncertain of the locations, drill small holes through the siding to find two or three stud centers. This will establish the stud pattern at 16-inch or 24-inch intervals. Hang a plumb bob and mark the stud bottoms below the siding.

2 **Checking for straightness.** Hold a straightedge such as a 1-by-4 against the wall vertically, horizontally and diagonally to locate bulges. If possible, pound and nail them into alignment.

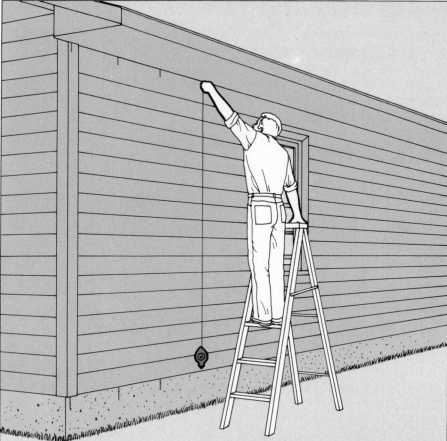

OLD POST

FURRING STRIPS   NEW POST

3 **Nailing the furring strips.** Attach 1-by-3-inch furring strips to each stud, using 2½-inch nails at 16-inch intervals to penetrate siding and sheathing. Drive the nails into the high spots of beveled siding. Where the wall bows, shim as necessary to keep the strips straight, checking frequently as in Step 2. Fur completely around windows and doors as well; if necessary, saw off projecting sill ends. If you do not plan to use corner boards with the new siding, nail furring strips at corners that do not have existing corner boards or adjacent to existing corner boards. If you plan to use corner boards, see Step 4. If you are re-siding with 4-by-8-foot plywood or hardboard panels, nail horizontal furring strips at the bottom and top of the wall, and at any intermediate point where rows of panels will join.

4 **Installing corner boards.** For outside corners, attach corner boards as thick as the furring strips plus the new siding, overlapping them at the corner. If corner boards already exist, nail the new ones directly over them. For inside corners (*inset*), nail a 2-by-2 strip, 1½ inches square, over the old post and the adjacent furring strips. Plane one edge of the new post, if necessary, to ensure a close fit.

Furring for Shingles and Vertical Siding

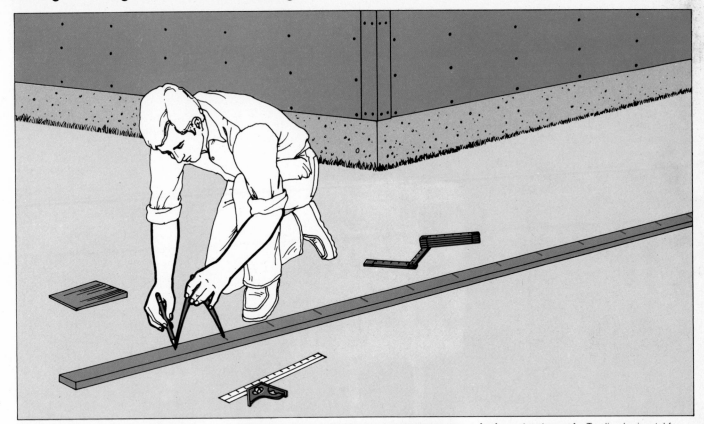

Laying out a story pole. To align horizontal furring, mark the intervals on a story pole—a straight piece of 1-by-2-inch wood long enough to cover the distance from the soffit to 2 inches below the existing siding. Set a pair of dividers at the height of a single piece of siding and mark the wide side of the pole evenly from top to bottom, squaring the lines with a combination square. Nail the pole tightly to the corners of the house and transfer the markings to the corner boards or furring strips. Snap chalk lines around the house to serve as nailing guides in mounting furring strips.

Furring out masonry. Using a story pole and chalk lines to establish the horizontal lines, nail 1-by-4-inch strips to masonry if you are siding with shingles. For vertical board siding, use 1-by-3-inch strips spaced 24 inches apart. Drive masonry nails directly into cinder block or into brick mortar joints at 16-inch intervals.

Retrimming Doors and Windows

The joints where new siding meets windows and doors can pose problems. If the old siding is so thick that the door- or window casing protrudes only slightly or not at all, you must protect the edges of the new siding by altering the casing.

Strips of molding nailed to the outer edges of the casing may bring the casing out as much as needed. If adding molding will not suffice, you can remove the casing, build out the jamb and install new casing (opposite). For windows and doors recessed into masonry walls, make a liner to cover the masonry (page 32).

Both extending jambs and lining masonry windows require ripsawing to exact widths. Use the method shown below.

A Jig for Accurate Ripping

The best way to make a dead-straight rip cut with a circular saw is to slide the edge of the saw's base plate along a true straightedge, which you can make by cutting a strip about 6 inches wide from the long side of a piece of plywood. Using nails, sandwich a 1-by-2 between the strip and a wider piece of plywood, so that the factory-milled edge of the strip overhangs the wide platform. Nail the platform to the top of a workbench. Slide the lumber under the overhanging straightedge; the distance between the rip line and the straightedge should equal that between the saw blade and the base-plate edge. Then tack two nails through the overhang to immobilize the lumber, and rip it with the saw set to cut slightly into the platform.

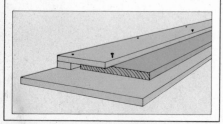

Building Out the Casing

Applying a molding. Bend the bottom edge of the existing drip cap out to the horizontal and nail the molding to the casing. If you use cove molding, as shown here, fasten it flush to the outer edge of the casing with nails driven through the narrow face at a slight inward angle. Nail backband molding (inset) from the side through the backband. Bevel the bottom ends of the side moldings to the slope of the sill. Miter the corners and set and fill the nails. Install a new drip cap (page 32, Step 4).

MOLDING

CASING

MOLDING

CASING

Building Out the Jamb

1 **Extending the jambs.** After gently prying off the old casing, nail ¾-inch extender strips flush with the outside edge of the jamb. The strips must be wide enough to project exactly the distance from the face of the jamb to the plane of the old siding. Butt side strips to the top one and saw the lower ends to the angle of the sill.

2 **Nailing on the casing.** Miter the top strip so that its bottom edge measures ¼ inch longer than the inside width of the new jamb, and nail it to the extender strip with the bottom edge ⅛ inch above the inner edge of the strip. Bevel the bottom ends of the side strips to the slope of the sill, miter the top ends to fit and nail them to the jambs ⅛ inch offset from the inner corner. Then set the nails and fill the holes.

EXTENDER STRIP

EDGE OF JAMB

EXTENDER STRIP

FACE OF JAMB

⅛" OFFSET

3 **Extending the sill.** Using a plane at least 10 inches long, plane the front edge of the sill flat and square. Then build the sill out 1½ inches with a strip of wood or of roundnose molding as thick as the sill.

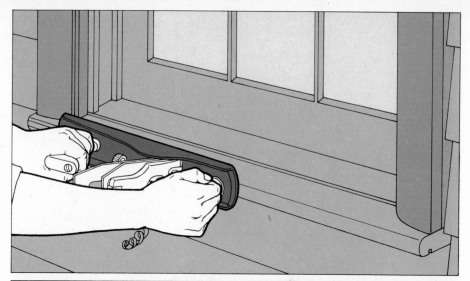

4 **Installing drip caps.** Shape a strip of flashing metal to overlap the front of the top casing strip ½ inch, cover the top of it and run 3 inches up the existing wall. Nail it at its top edge only.

Lining a Recess in Masonry

Cutting the liner boards. Measure from the casing out to the front of the furring strips that will support the new siding. To this measurement add the thickness of the new siding plus ½ inch to bring the liner out slightly beyond the siding. Rip ¾-inch boards to this width. Saw to length so that the side boards will butt against the top board and fit the angle of the sill. Glue the boards in place with exterior panel adhesive applied generously. Caulk the joints where the liner meets the casing and the sill.

Precut Panel Siding that Covers a Wall Fast

For covering a wall quickly and durably, panels of plywood or hardboard are hard to equal. They can be nailed directly over sound clapboard or shingles, and because they are stiff they will flatten small irregularities in such surfaces. Most are designed to be installed vertically and are made 4 feet wide and long enough to extend from the foundation to the eaves of one-story houses. Standard thicknesses—5/8 inch for plywood, 7/16 inch for hardboard—are best.

In one common style, 3/8-inch vertical grooves spaced 8 inches apart give the appearance of boards backed by battens. Another has a rough surface resembling wood fresh from the saw, and another a high-relief grain made by wire brushing. Hardboard is also molded into simulations of stucco, shakes or shingles. Even plain hardboard can be patterned with vertical battens.

Panels should overlap the foundation 2 inches; allow an expansion gap of 3/16

inch where panels meet the soffits, rake boards and door- and window casings. When the height of a wall exceeds the length of the panels, as in a gable wall or a two-story house, install a second course above the first. Locate the plate that lies atop the studs and cut panels so that the horizontal seam lies over the center line of the plate, and nail them into it.

Seal all edges. Fasten panels with galvanized nails, preferably long enough to go at least 1 inch into the studs.

BACK-LAP EDGE

1 Positioning the first panel. Set a panel, cut to length, to overlap the corner of a wall, with the top snugged up to the soffit and the back-lap edge ¼ inch short of the center line of a stud; mark the corner edge in line with the thickest parts of the clapboards or shingles of the adjoining wall.

Cut the panel at the mark and start a nail midway on each long edge at the stud positions.

Set the panel back in place with a ³⁄₁₆-inch spacer stick—a yardstick will do—between it and the soffit; then, while a helper secures the

panel by leaning into it hard, drive the nails home. Drive additional nails every 6 inches around the top, bottom and corner edges and every 9 inches along the intermediate studs, but not into the back-lap edge; drive each nailhead just to the surface of the panel. Remove the spacer stick.

2 **Joining panels.** Install each succeeding panel with its front lap over the back lap of the preceding panel; align the seam between two panels at the top and bottom, and fasten the new panel in place. At joints do not nail through laps; instead drive nails through the body of the new panel (inset), leaving the adjoining panel free to expand and contract.

3 **Fitting around windows and doors.** Cut the siding to fit around door- and window casings with a 3/16-inch gap on all sides for caulking. To retain lapped outer edges, cut lengths from full-width panels rather than remnants.

To make the inside cut shown below with a circular saw, pull the guard out of the way, line the blade up over the cut mark with the forward edge of the base plate resting on the panel and the blade slightly above the panel. Start the saw and using the base-plate edge like a hinge, slowly lower the blade into the panel. Saw to the marked corners and cut out the small remaining arc of material with a handsaw.

SPACER

SHEATHING STUD
OLD SIDING BACK LAP
PANEL BODY FRONT LAP

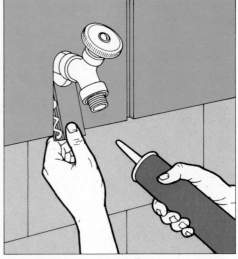

4 **Slotting the panel for a pipe.** Where a pipe or utility cable must pass through the siding and cannot be disconnected and later passed through a hole, mark the point by measuring the vertical distance from the bottom edge-line of the panel and the horizontal distance from the edge of the adjacent panel. Drill a hole slightly larger than the pipe and bevel-cut toward it so that the slot cut is slightly wider at the front than at the back. With a plane, bevel the sides of a strip of panel material so that you can push it in just flush to the surface. Apply panel adhesive to the sides of the filler strip and press it into place.

5 **Installing a course in a gable wall.** Use a plumb bob to mark points on the rake board or soffit that are directly above the joints. Measure from the joints to the marked points to find the edge dimensions of angle-topped panels to be fastened to the gable wall.

To waterproof the joint between the upper and lower panels, double-bend strips of flashing metal 3 inches wide so that the cross section roughly resembles a Z. Tack the strips in place so that one bent edge extends down in front of the top of the lower panels and the other will lie in back of the bottom of the upper panels (*inset*). End-lap the strips 3 inches. Nail on the panels.

6 **Capping outside corners.** Make corner boards by nailing a 1-by-3 on one side of the corner, then overlap with a 1-by-4 on the other; the direction of the overlap should duplicate that of the panels. Fasten each strip to the panels with staggered nails; do not nail the strips to each other. Lay a heavy bead of caulking on the end grain of any board with an exposed top, shaping the caulking to drain rainfall.

1 × 3 — — 1 × 4

7 Capping inside corners. Install inside cove molding *(inset)* to protect the joint where the panels meet. Fasten through both edges of the molding, using galvanized siding nails and butting the top of the molding to the soffit.

COVE MOLDING

8 Installing a frieze board. Seal the joint where the top edge of the siding meets the soffit with a new frieze board. Cut 1-by-6s into lengths so the joints will lie over studs. Use bevel joints where the boards meet, and butt joints against the corner boards and inside cove moldings.

9 Protecting the top edge. If the rake board does not cover the upper edge of the siding and the gable has no overhang, double the board with a strip of lumber about 1½ inches wide and thick enough to cover the exposed edge. Seal the strip with caulking between the strip, the rake board and the siding edge, and caulk any butt joints along the length of the strip. If the overhang is deep enough, run the siding up to it and nail on a new rake board in the same way that you fastened the frieze boards.

Lapped, Locked or Battened—the Versatile Board

A few rotting clapboards can turn a handsome house into an eyesore. In many cases, the offensive boards can be cut out and replaced (page 38). But if you decide to apply an entirely new skin of wood siding, you have a wide choice.

Clapboards may be plain or beveled. Other styles of wood siding boards have interlocking edges, and still others make a flat surface. Many types can be applied either horizontally or vertically, while some are suitable only for one or the other. Siding ranges up to 1 inch in thickness, 12 inches in width and 20 feet in length. Though cedar is the most common siding wood, boards are available in redwood and in the man-made wood composition, hardboard.

Aside from the type of siding to use and whether to apply it vertically or horizontally, the only major consideration before starting work is how to finish corners. With horizontal siding, one option is to use corner boards (page 28). Alterna-tively you can weave clapboards like cedar shakes or shingles (page 52). Or you can substitute metal corners for outside corner boards. These corners (page 40) simulate the effect obtained by mitering the ends of siding boards, a procedure too demanding for all but a seasoned carpenter. Vertical siding is applied much like horizontal siding except for a few details (page 41).

Siding must be fastened to a sound surface. Existing siding that is badly damaged must be removed. Peeling paint may be a sign of moisture problems to be corrected. Window frames, doorframes and trim boards usually have to be built up to accommodate new siding (pages 30-31). Downspouts and other wall fixtures must be removed. Furring strips are a necessity when re-siding over masonry or applying new horizontal siding over old clapboards. Horizontal siding on top of a flat wood surface, such as tongue-and-groove siding or vertically applied siding, requires no furring strips unless the surface of the wall is uneven.

Nails should be stainless-steel siding nails with spiral or annular ring shanks for the tightest grip. Follow the nailing patterns below and overleaf, and when fastening the end of a board, especially a thin beveled one, drill a small pilot hole for the nail to prevent splitting.

Before nailing the first board, determine whether the old siding is level. If it consists of straight boards or panels, lay a level against the bottom edge of the siding on each side of the house. If the old siding is level, use it as a guide for the new. If the courses are canted or if the siding has an irregular edge, measure the house to be sure that the old siding will not show below the new (page 39).

For vertical siding, check not only that the old siding is level, but that the corners of the house are plumb. If they are not, the first boards must be shaped to correct the lean (page 41).

How to Nail the Joints

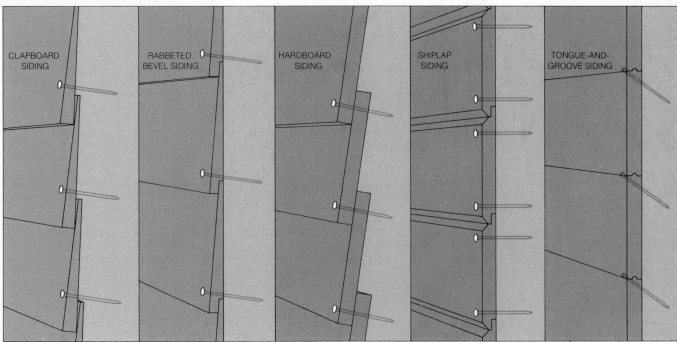

Horizontal siding. Clapboard and rabbeted bevel siding closely resemble each other and are both nailed to studs or furring strips so that the nail misses the board below. Use one eight-penny nail in each stud or furring strip for ¾-inch-thick siding, one sixpenny nail for thin-ner boards. Hardboard siding, less likely to split, is nailed through the piece below. Shiplap sidings wider than 6 inches are face nailed with two eightpenny nails, hammered one quarter of the board width from the edge. With narrower shiplap the upper nail is unnecessary. Tongue-and-groove siding 6 inches wide or less is blind nailed—fastened at the base of the tongue with a 45° nail so that the head is hidden by the board above—with a single eightpenny finishing nail. Countersink the nail with a nail set. Wider boards are face nailed like shiplap siding.

Vertical siding. In the board-and-batten style, strips called battens cover joints between the wider boards. Fasten the boards first. Space them ½ inch apart and hammer eightpenny nails through the center of each board at 24-inch intervals. Cover the gaps between boards with battens at least 1½ inches wide, nailed through the center every 24 inches with a tenpenny nail.

Board-on-board siding, similar to board and batten, has underboards spaced to allow a 1½-inch overlap by the top boards at both edges and nailed through the center with eightpenny nails. Top boards are attached with tenpenny nails driven next to the underboards.

Channel siding is a variety of shiplap used only vertically. Face-nail it with two eightpenny nails spaced to miss the rabbets in the board edges by ½ inch.

In addition to these styles, shiplap and tongue-and-groove sidings can also be applied vertically with the nailing patterns shown on page 37.

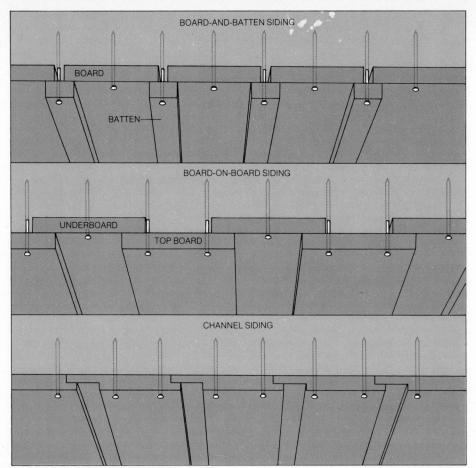

Replacing a Damaged Board

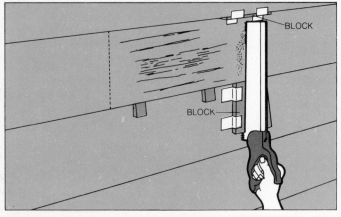

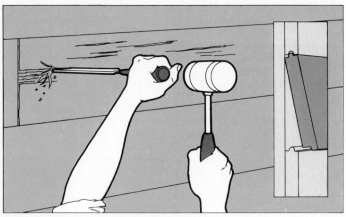

Repairing clapboard siding. To replace a short section of damaged clapboard, tap wedges under the board to raise it. Tape protective blocks of wood to the edge of the board above and to the face of the one below, and with a backsaw cut through the exposed part of the raised board. Insert the wedges under the overlapping board and complete the cuts with a keyhole saw. Pull out the damaged section, removing any nails you find. Use a hacksaw to cut hidden nails (page 48). Fit a replacement section into position and nail it as shown on page 37.

Repairing tongue-and-groove or shiplap. Chisel through the board, first vertically across it outside the damaged area, then along a 1-inch channel across the middle. Pry away the damaged pieces and cut a new section to fit. Shiplap siding slides easily into place. With tongue-and-groove siding, chisel away the back side of the groove (*inset*) and slip the board into place, tongue first. Nail the board through the face to secure it.

Installing Horizontal Boards

1 Measuring for a level first course. If the old siding is not level—check with a level along the bottom—record the height of the siding at each corner of the house; measure from the fascia or soffit to the bottom edge of the existing siding. Transfer the longest measurement to each corner, then snap a horizontal chalk line between the marks around the house.

Between the chalk lines, nail vertical furring strips to the old siding and add new corner boards *(page 28)*, unless you plan to install metal corners *(page 40)* or weave clapboard ends as shown for shingles on page 52. Extend window and door casings *(pages 30-31)* and gable-ends *(page 35)*, if necessary, to accommodate the new siding.

2 Nailing the starter strip. For clapboards, nail a starter strip over the furring strips, flush with their bottom edges. Use a board about 1½ inches wide and as thick as the top edge of the siding, and fasten it to each furring strip with a sixpenny nail.

Rabbeted-bevel, shiplap and tongue-and-groove sidings require no starter strips.

3 **Applying the siding.** For the first course, start a nail near the end of the board so the nails will enter furring strips or studs. With a helper, hold the board so that it overlaps the bottom edge of the starter strip by ⅛ inch or, where no starter strip is used, the joint between sheathing and foundation by 1 inch. If you use corner boards, be sure to make tight joints between them and siding. Butt lengths of siding at furring strips or studs, staggering the joints for weathertightness.

Subsequent courses of rabbeted-bevel, shiplap and tongue-and-groove siding space themselves. For clapboard siding, make a story pole (*page 49*) with divisions the width of the exposure. Use the pole to mark corner boards, then snap a chalk line for each course.

(*page 49*)

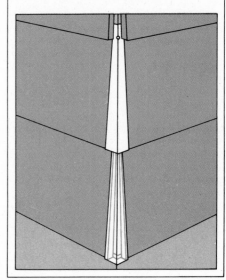

A Laborsaving Corner

The finely crafted appearance of mitered corners can be approximated with clapboard siding by the use of special metal cornerpieces over the ends of boards that have been nailed flush with the corners of the house. (Similar metal corners are available for the asbestos shingles shown on pages 62-63.) Used in place of outside corner boards (*page 28*), the metal corners slip onto the ends of each course after all the siding on adjacent walls has been applied, and then are nailed to the corner of the house.

pages 62-63. (*page 28*)

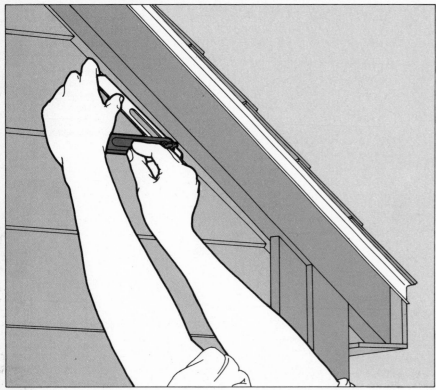

4 **Fitting at obstacles.** At gables, measure the rake angle with a T bevel and cut the ends of the siding to match. Trim the width of the topmost course if necessary. Caulk all joints between the ends of siding boards and the house.

To fit siding around corners of windows and doors, saw away part of the board to leave a tongue that fits snugly into the space above or below the opening. Do not drive nails into tongues narrower than 2 inches.

Installing Vertical Boards

1 **Installing the first board.** If you find that the corners of the house are not plumb, plane or rip an edge of the first board so that when nailed in place, one edge will be flush with the corner or corner board and the other edge vertical (exaggerated for clarity below). Align the bottoms of boards with a chalk line around the house *(page 39)* and nail them in place, using the nailing patterns shown on page 38.

2 **Butting vertical boards.** To help keep water from seeping into a joint, bevel the ends of adjoining boards before nailing either to the wall. Stagger joints for a pleasing appearance.

CORNER BOARDS

FIRST BOARD

3 **Beveling for a tight fit.** When fitting a shiplap or tongue-and-groove board along a window or doorframe, rip the board to fit the space, then bevel the edge with a plane to the shape shown at left. Insert the tongue into the groove of the preceding board, then push the beveled edge against the wall. Nail the board in place.

If the board is cut out to fit around the frame, use a chisel to complete any part of the bevel that a plane cannot cut.

Metal and Plastic—Boards without Wood

Low cost and ease of installation make lightweight vinyl and aluminum siding the most popular materials for re-siding jobs, and increasingly for the outer covering of new walls. Both are available with insulation backing if desired.

Choosing between the two, which are remarkably similar in panel sizes and shapes, mounting techniques and durability, is mostly a matter of taste and availability. Vinyl is a bit easier to work, does not dent or show scratches and resists heat and cold better than aluminum. But it can crack when struck in cold weather and cannot be repainted.

Aluminum is usually less costly than vinyl, and comes in a brighter palette of factory-baked enamel finishes. But it is easily dented and scratched, and some codes require electrical grounding.

Whichever material you use, the key to professional-looking work is in the installation of the accessories—F channel, J channel, corner posts, starter strips and trim. When these are mounted level or plumb in their correct locations, the rest of the siding job is mostly routine sawing and hammering. Aluminum requires no special tools, but you will need a snap-lock punch, available at hardware stores and siding dealers, to indent the edges of some vinyl panels *(page 44)*. When replacing vinyl panels, you will also need a handy zipper tool *(above, right)*.

Both vinyl and aluminum siding expand and contract more than wood, and allowances must be made for this movement. Always provide a ¼- to ⅛-inch space at joints and panel ends. Never pull vinyl panels taut before nailing, since this may cause dimpling or rippling as they expand. With either material, drive nails straight into the wall through the centers of the oval nailing slots, leaving a gap of about ⅟₃₂ of an inch between the nailheads and the material—the thickness of a matchbook cover.

While it is possible to apply vinyl or aluminum siding without installing a new soffit—or vice versa—it is usually advisable to do both. If you already have a soffit, you may simply want to cover it

with new material. If you do not have one—or choose to rip out the old one—use at least one perforated panel *(page 46)* every 10 feet for ventilation.

Like wood, vinyl and aluminum siding can be installed horizontally, vertically or—by special arrangements of mounting appliances—in both directions on the same wall. To make the lines of overlapping joints virtually invisible, begin siding at a point farthest from the usual viewing locations—front sidewalk, front entrance, driveway—and work toward those locations, lapping the previous panel. The finished wall will appear to be an unbroken expanse of siding when viewed from the "favored" side.

A Zipper and Cement for Replacements

Replacing a vinyl panel. Use a special tool called a zipper, available from siding dealers, to reach under the undamaged upper panel in order to hook into the locking strip inside the top of the damaged panel *(below)*. Pull down firmly while sliding the zipper the length of the damaged panel. This will unlock the upper panel, which can then be propped up high enough to remove the nails holding the damaged panel. Lock and nail the replacement panel, then use the zipper to relock the upper panel.

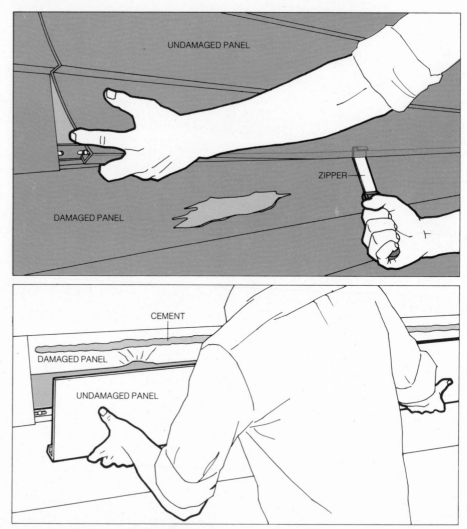

Replacing an aluminum panel. Use a utility knife to cut a slit along the center of the damaged panel, or just above the center groove on a panel that simulates a double row of siding. Unhook and discard the lower portion, leaving the upper portion intact. Cut off the nailing and locking strip of the replacement panel. Apply roofing cement the length of the damaged half-panel. Install the new panel by sliding its top edge under the locking hook of the panel above, while locking it into the lip of the panel below. Press down firmly to cement the panel.

A Horizontal Layout

1 **Installing a soffit track.** If you plan to include a new soffit with the siding, begin by mounting a track of F channel to receive the soffit panels (*page 46*) at the top edge of the siding. Use a level to mark points on the wall that line up with the bottom of the fascia board, then strike chalk lines between those points. If you are using vinyl, nail facing strips of F channel both to the bottom of the fascia and to the wall; with aluminum, only the wall channel is needed.

If you are not adding a soffit, simply nail strips of undersill or general-purpose trim (*page 44*) along the upper edge of the existing siding.

2 **Nailing in corner posts.** To provide vertical support for horizontal panels, attach grooved corner posts the height of the siding, setting them ¼ inch below the top trim. To make sure they are vertical, pin each with a couple of nails at the center and check the plumb (*below*) before nailing. Drive nails at 12-inch intervals along the nailing flanges (*inset*) on both sides.

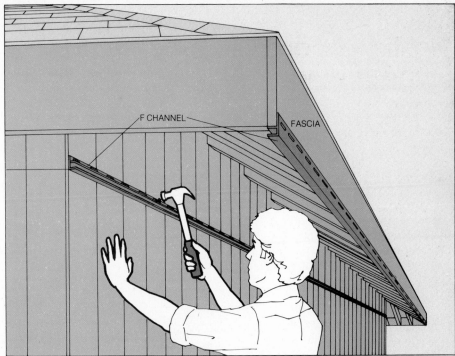

3 **Mounting window and door trim.** Strips of J channel, mitered at the corners, are nailed to the tops and sides of all large wall openings—windows, doors, louvers, air-conditioner recesses. To allow for mitering, cut the top piece longer than the top by two channel widths, and the side pieces longer than the sides by one channel width. Nail the top piece first, then carefully fit and nail the side pieces (*left*).

4 **Installing the panels.** Lock the first panel into the lip of a starter strip, nailed along the bottom of the old siding, before sliding it into the corner post (*below*) and nailing it every 16 inches. After the starter course, lock each panel into the one below. Where a panel meets a corner, allow ¼ inch for expansion.

5 **End-to-end joints.** To join panels on the same course, lap the factory-notched end of one panel by about an inch with the unnotched end of the next, leaving a gap between nailing strips. If you use aluminum, slip a small metal backer plate behind each joint for support, and nail it to the wall before nailing the panel. Stagger joints in adjoining courses by at least 4 feet.

6 **Cutting around wall openings.** When there is less than a panel width between the top of a course and bottom of a window or other opening, notch a panel to fit snugly underneath. Hold the panel against the opening to mark the cutout dimensions, adding ¼ inch in each direction for expansion. Make vertical cuts with metal shears (*below*); score the horizontal

cut with a utility knife and snap out the section. If you are using insulation-backed panels, cut the insulation before scoring.

To secure the cutout section under a window, nail a strip of undersill or general-purpose trim beneath the sill. Add furring if necessary to maintain the pitch of the siding when the panel is locked into the trim (*inset*).

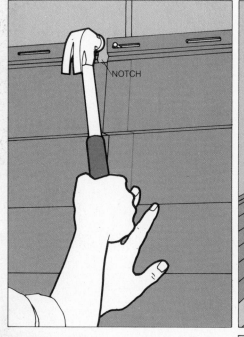

7 **Installing the top panel.** The top course is not nailed, but is tucked into a strip of undersill trim. After the trim has been nailed just beneath the soffit track, or along the top of the existing siding, measure the gap between the last course and the trim. Cut the final panels to fit, and use a snap-lock punch (*inset*) to indent the upper edge of each panel every 6 inches. Then push the panel into the trim while locking its bottom edge into the lip of the panel below. The indentations will lock the top into the trim.

A Vertical Layout

1 **Mounting the starter strip.** If you are using vertical vinyl panels, start from the center of the wall. After you have nailed in corner posts, mount strips of drip cap along the bottom of the wall, and J channel along the top. Then drop a plumb line from the center of the wall to mark the location of the double-channeled starter strip. Nail the strip precisely on the plumb marks, allowing ⅜ inch at the top for expansion.

If you are using aluminum instead of vinyl, install J channel at both top and bottom; omit the corner posts and starter strip.

2 **Mounting the panels.** After cutting panels to fit between trim strips, allowing ⅜ inch for expansion, install each one by inserting it into the top strip (*below*) and then lowering it onto the bottom strip before locking it to the adjoining panel. With vinyl, work from the center until you are a panel width or less from the corners.

With aluminum, begin about a foot from the end of the wall. Cut off the simulated batten at the outer edge of the starter panel and plumb carefully before nailing on both sides. Then proceed across the wall.

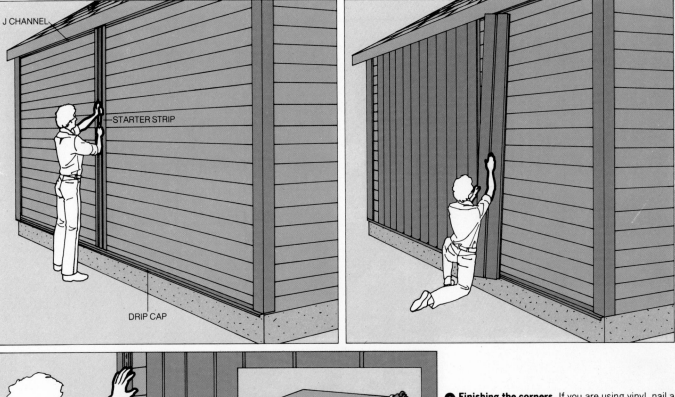

3 **Finishing the corners.** If you are using vinyl, nail a piece of undersill trim inside the corner post (*left*), furring if necessary to keep it straight. Do not nail the trim through a post nailing flange. Cut the last panel to fit, allowing about ¼ inch for expansion and indenting its edge with a snap-lock punch so it will lock into the corner trim.

With aluminum, fit the last panel on each wall so that it extends past the corner and mark points ⅛ inch beyond the corner at top and bottom. Remove the panel, clamp it between boards and use another board to bend it to a 90° angle at the marks (*inset*). Remount the bent panel and nail it around the corner. When you reach the last corner and return to the starting point, mount a strip of general-purpose trim over the nailheads along the outer edge of the starter panel to receive the last corner panel.

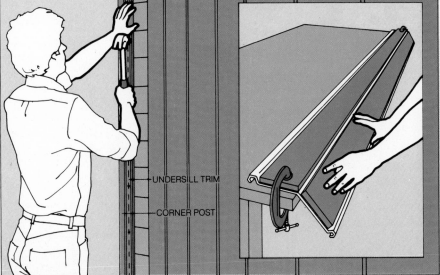

Mounting a Soffit

1 **Installing soffit panels.** Soffits are made of specially constructed panels, their sides locked into one another and their ends fitted into tracks. For vinyl *(below, left)*, cut the panels to fit between the channels of the soffit track *(page 43)*, allowing ¼ inch for expansion. If the span is greater than 18 inches, hang nailing strips, 2-by-2 inches, from the rafters to the soffit midpoint. Bend the panels slightly to slip them into the ends of the track and lock them together while you push them toward the center of the soffit run. If you have a nailing strip, use a pair of aluminum trim nails to fasten each panel to it. For aluminum *(below, right)*, where only one mounting channel is used, cut the panels to cover three fourths of the fascia bottom, leaving space for the fascia cover *(Step 3)*. Set one end of each panel into the channel, allowing ⅛ inch for expansion. Make sure it is square and locked into the preceding panel, then nail the other end to the fascia with aluminum nails.

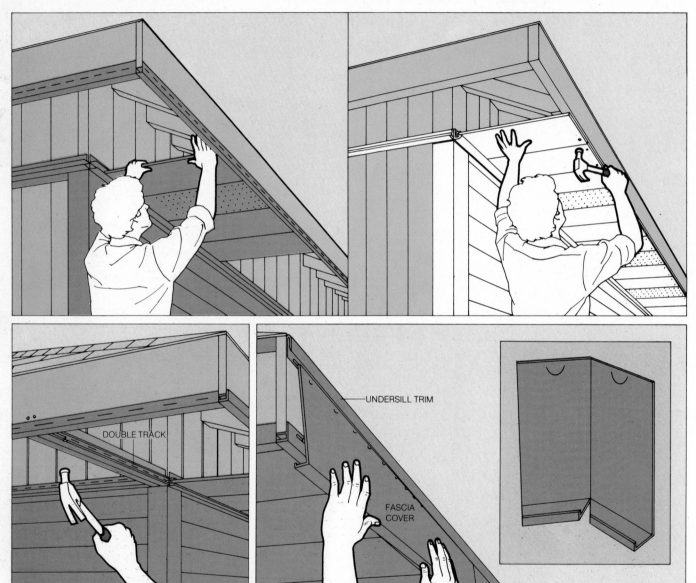

DOUBLE TRACK

UNDERSILL TRIM

FASCIA COVER

2 **Finishing a soffit corner.** When two runs of soffit panels meet at a corner, a simple channel extension permits one run to fill the corner gap. Nail a double track to the bottom of a 2-by-2 strip nailed between the wall and fascia, parallel to one row of soffit panels and perpendicular to the other. Install panels to fill the two runs, trimming and bending the last panels to fit securely.

3 **Covering the fascia.** Finish the soffit installation by covering the fascia board with special J-shaped panels. If you are using vinyl, nail a strip of undersill trim along the top of the fascia, cut fascia covers to the correct width and indent the top edges with a snap-lock punch *(page 44)*. Install the covers by hooking them over the soffit channel and locking them into the trim *(above)*. Make a corner cap *(inset)* by cutting off a 5½-inch piece of fascia cover, marking and lightly scoring a vertical center line on the back, and cutting out a 90° section of the flange. When indented and bent, the cap will fit snugly. If you use aluminum, nail the covering to the fascia top, where the nailheads will be concealed by gutters or an L-shaped trim.

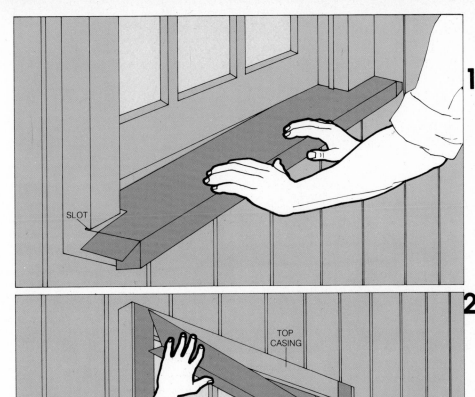

A Vinyl Cover for Window Frames

1 **Covering the sill.** Although window sills and casings are usually left uncovered, some manufacturers provide optional vinyl coverings that can be installed before mounting window trim (*page 43*). Before covering the sill, trim the ends flush with the outside edges of the casings and saw a slot about 1/16 inch wide—the width of a keyhole-saw blade— and 1/8 inch deep at the bottom of both casings. Cut the cover big enough to overlap the sill 2 inches at each end, and apply vinyl cement on the back and the sill. Press the cover onto the sill, slipping it into the slots before folding the ends. Seal the folds with single nails, and nail every 6 inches beneath the sill.

2 **Covering the casings.** The L-shaped coverings for the casings are fitted first onto the sides and then to the top. Cut the side casings to extend all the way from the sill to the upper edge of the top casing, notching the inside edge to pass over the top casing. Fit the covering onto the casing; nail the inner edge every 6 inches. Measure and cut the top casing long enough to extend across both side casings. Then cut out a 45° angle at each end, so that the angles lap the top of the side casings (*left*). Nail the covering along the inner edge.

Corner Framing for Uneven Soffits

Boxing in gable ends. On some gabled houses two runs of soffit meet unevenly at the corners, and you may have to build a frame of 2-by-2s nailed into the fascia and corner studs. Nail strips of F channel to provide an extension for the eave soffit and J channel to hold the pieces of fascia cover that box in the frame.

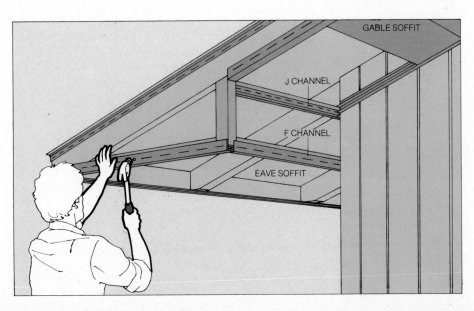

The Traditional Beauty of Shingles and Shakes

Western red cedar or redwood shingles and shakes vary in texture, thickness and length, but all make sturdy wall coverings that go up easily. Although they can be painted or stained—and some come already finished—cedar weathers naturally to a handsome silver gray; redwood, however, turns black if left unfinished. Wood withstands the elements, but if damage occurs, individual shingles or shakes can be readily removed with a chisel. Old nails usually are hidden by overlapping shingles or shakes (below); new nails for a replacement are small-headed ones that can be driven into the face and left exposed.

Shingles and "grooved shakes" are machine-made tapered slices; regular shakes are hand-split uniform slices as much as 1½ inches thick. All are installed in overlapping rows, or courses.

Single coursing (pages 50-52) sets rows close together, with the bottom half of one course overlapping the top half of another. In double coursing (page 53), each shingle or shake is installed over an inexpensive backer, of the same length, called an undercourse shingle, so the rows can be set farther apart. Grooved shakes are always double-coursed.

For single coursing, the maximum exposure between rows is ½ inch less than half the length of the shingle or shake. For double coursing, maximum exposure is usually 4 inches less than the full length. To determine exposure, measure the major wall of your house from the soffit down to 1 inch below the top of the foundation. Divide the measurement by the desired exposure, then round off the result to find the courses you need. To get evenly spaced rows, find the exact length for each exposure by dividing the number of courses into the wall measurement. Then make a "story pole" (opposite) to see how the courses will work.

How many bundles of shakes or shingles you need will vary with exposure, the coursing method and the dimensions of the house. As a rule of thumb, you can figure four bundles of shingles or five of shakes for each 100 square feet of wall—and about 10 per cent extra for trimming and fitting.

Shingles or shakes require a wood nailing surface. If your wall is clapboard or smooth plywood, nail directly to the old siding. Because shingles or shakes are nailed near the middle, not at the top, they lie flat even against lapped siding. With board-and-batten siding, however, you must rip off protruding battens. If you have mineral fiber shingles, remove them (page 26). And if your siding is hardboard, brick or stucco, put up 1-by-4 furring strips (page 29), using a story pole to determine spacing between strips.

Putting new siding over old may create such a thick cover that window and door casings look buried in the wall. If you need to remove or extend old casings, prepare them in advance (pages 30-31). Before re-siding, fill the inside corners of the walls with 2-by-2 molding strips such as those used for clapboard siding (page 28, Step 4). If you are using shingles, outside corners can be covered ahead of time with corner strips 1 inch thick (page 28). With either shingles or shakes, outside corners can be finished as the siding is applied by weaving the edges as shown on page 52—a method unique to these materials. After the walls are done, caulk seams at casings or other trim.

Fitting a Precise Patch

1 Removing the broken unit. With a wood chisel and hammer, split the shingle or shake along the vertical grain lines, then pull out and discard all the pieces. Select a replacement shingle or shake ¼ inch narrower than the gap left in the siding, or use a hatchet to cut a new shingle or shake to the required width.

2 Taking out old nails. Pull nails that protrude near the bottom of the gap. Or, if no nails are visible, use a hacksaw (above) to cut off the tops of the first set of nails covered by the shingles or shakes above the gap. In most installations, the hidden nails will be about 1 inch above the butt of the covering shingle or shake.

3 Installing the replacement. Slide the new shingle or shake as far into the gap as possible. Measure from its butt to the butts of the adjoining shingles or shakes. Cut this amount from its tail end, replace the new unit and drive small-headed galvanized sixpenny nails 2 inches above the butt and ¾ inch in from each side.

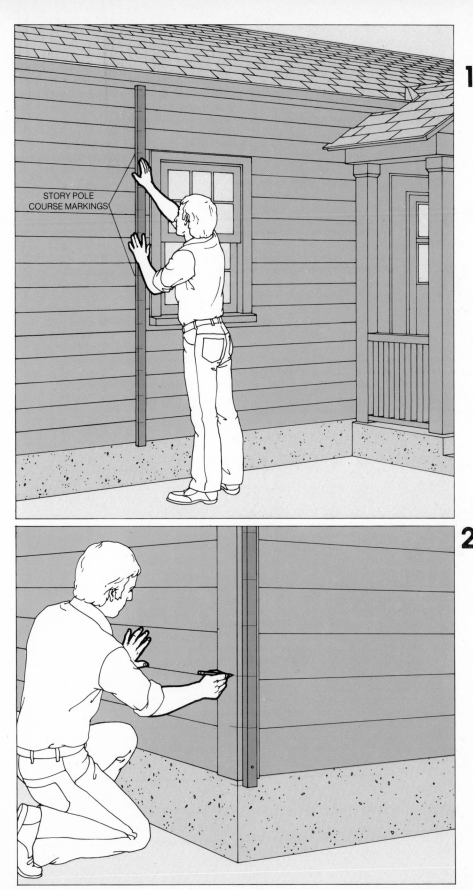

STORY POLE
COURSE MARKINGS

Planning the Courses

1 Making the story pole. To determine the ideal starting point for each course of shingles or shakes, use a story pole. Make one from a 1-by-2 the height of the major wall of your house—measured from the soffit to 1 inch below the existing siding or sheathing. In a two-story house, you will have to splice pieces to make a long story pole. Draw marks on the wood at evenly spaced intervals equal to the desired exposure for each course. Then hold the story pole alongside a window casing with the top of the pole jammed against the soffit. Adjust the spacing marks if necessary so courses above and below the casing will be at least 4 inches deep.

2 Laying out the courses. Holding the story pole tight against the soffit—and tacking it in place if necessary—transfer the pole markings to both sides of each inside and outside corner, and to all window and door casings. Make sure the marks at the bottom level of the pole are 1 inch below the top of the masonry foundation. Then drive a hardened masonry nail partway into each corner of the foundation at the bottom marks on the major wall of your house. Stretch a string between the nails to indicate the butt line for the starter course of shingles or shakes.

The Shingler's Hatchet

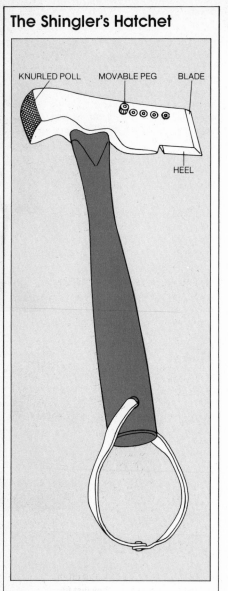

KNURLED POLL   MOVABLE PEG   BLADE

HEEL

This many-purpose tool serves as a hammer, a utility knife and a measuring gauge as well as a hatchet for splitting wood shingles or shakes to size. The poll is knurled to keep the hammering end from slipping off nailheads in high-speed nailing. The sharp heel of the blade can be drawn, like a knife, along an edge to shape an angle. The notch behind the heel can be used for pulling nails. The movable peg and predrilled holes form a gauge to measure—from poll to peg—distances of 3½ to 5½ inches.

Single Courses for Shingles or Shakes

1 **Beginning the starter course.** Hold the first shingle or shake upright with the thick butt end touching the string across the foundation. Set the outside edge of the shingle or shake flush with the outside corner of the wall. Drive threepenny nails ¾ inch in from the side edges and 1 inch farther from the butt than the desired exposure for the shingle or shake. Add a nail in the middle if the shingle or shake is more than 8 inches wide.

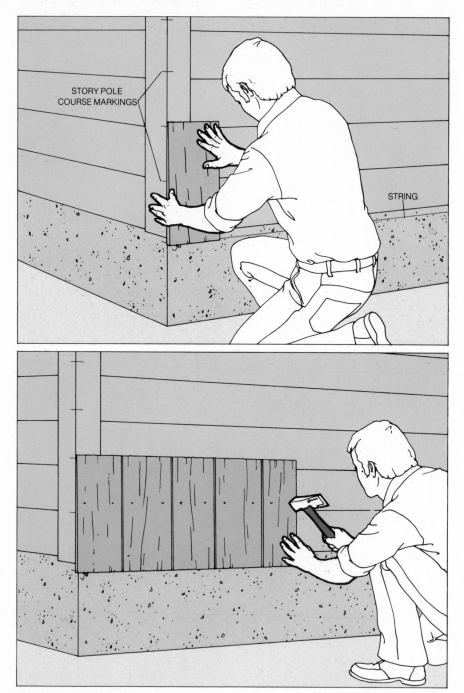

STORY POLE
COURSE MARKINGS

STRING

2 **Finishing the undercourse.** Work along the string to the far corner of the wall, spacing the shingles or shakes ⅛ inch apart unless they are "rebutted and rejointed" shingles that can be fitted snugly together. Use a shingler's hatchet to trim the last shingle or shake so that its outside edge will fit flush with the edge of the far corner, or sort through the random widths in the bundle to find a shingle or shake of the required dimensions.

3 **Completing the starter course.** Starting at the original corner again, install another layer of shingles or shakes over the undercourse layer. Align the butt ends, but let the outside edge of the corner shingle or shake extend 1½ inches beyond the edge of the outside corner. Work across the row, offsetting the joints between shingles or shakes so that those in the outer layer fall at least 1½ inches from the joints in the undercourse. If the string ends at an inside corner, butt the outside edge of the last shingle or shake against the corner strip. If the string ends at an outside corner, let the last shingle or shake extend at least 1½ inches over the edge.

4 **Installing successive courses.** Tack a 1-by-4 over the first course with its top edge at the level indicated for the next course by the story-pole marking, and line up the butts of that course against it. Select for the first shingle a width that will offset the vertical joints by at least 1½ inches from those in the preceding course, while extending at least 1½ inches beyond the outside corner. For each successive course repeat the process, moving the board up each time. Use a hatchet to split shingles or shakes to size at window casings or other trim.

5 **Installing the soffit course.** Cut off the tops of the shingles or shakes so they just fit, and nail them ¾ inch in from each side edge and about ¾ inch above the butt. Cover the crack at the soffit with molding.

6 Starting the woven corner. Nail on the under-course layer of shingles or shakes, using the methods on page 50, Steps 1 and 2. Hold the first shingle or shake for the outer layer of the starter course in place and use it to draw a line down the back of the shingle or shake extend-ing from the adjoining wall. With a hatchet, trim the extending piece along the line. Nail the unit in place, butting the outside against the back of the trimmed piece. Finish the course, letting the final shingle or shake extend 1½ inches be-yond the edge of an outside corner.

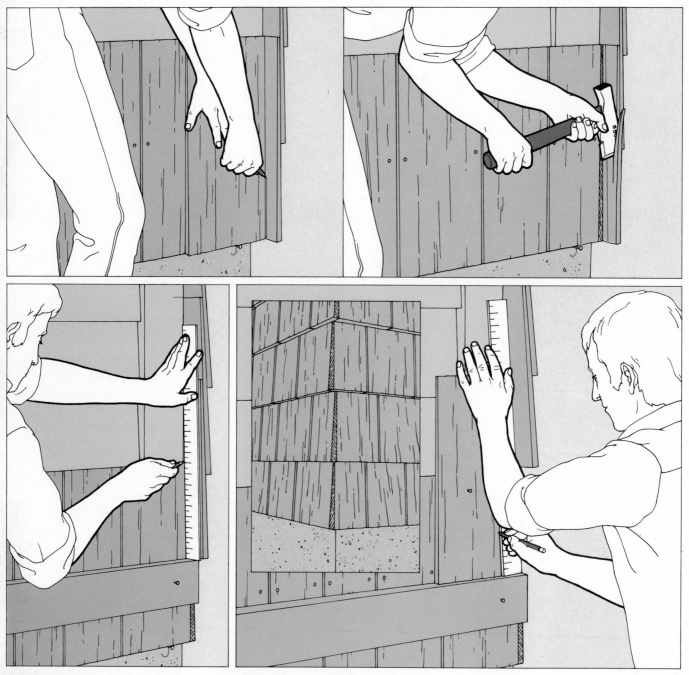

7 Alternating the weave. Tack the 1-by-4 across the installed shingles or shakes at the level indicated by the story pole course markings. Then mark the back of the extending unit of the adjoining course by drawing a line level with the face of the installed shingle or shake. Trim the extending edge along the line, taking care not to cut the overlapping course.

8 Finishing the woven corner. Tack a shingle or shake in place at the corner. Hold a straight-edge upright at the outside edge of the shingle or shake, aligning its outer side with the back of the shingle or shake in the adjoining course. Draw a line along the inner side of the straightedge, remove the marked unit and saw off the outside section on the diagonal line. Then fit the sawed edge of the new unit over the trimmed edge of the shingle or shake in the adjoining course and nail. If necessary, shave the unsawed edge of the unit to offset it from the vertical joint in the preceding course. Complete the course. Continue installing shingles or shakes up the wall, trimming units extending from the ad-joining courses to weave the corner (inset).

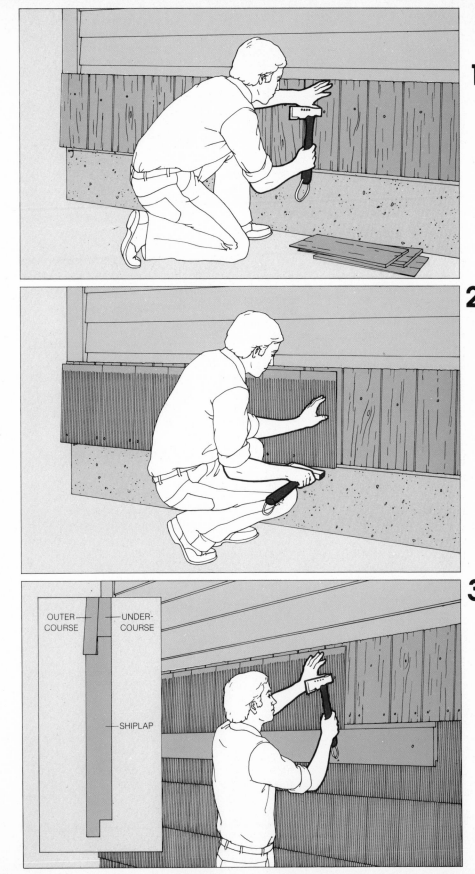

Double Courses
with Backers

1 Undercourse. Starting flush with an outside corner, install low-grade undercourse shingles across the wall, spacing them about ⅛ inch apart and keeping them about ½ inch above the string suspended on the foundation. Secure each shingle by driving a single sixpenny nail through the thin tail. Cover the row with a second layer of undercourse shingles, staggering the joints at least 1½ inches between the two layers.

2 Installing the first course. Hold an outer shingle or shake upright with the butt touching the string and the outside edge extending 1½ inches beyond the outside corner. Drive a fivepenny nail ¾ inch from each side edge and 2 inches above the butt. If the shingle is 8 or more inches wide, drive another nail midway across the shingle 2 inches from the butt. Work along the string to the opposite corner, laying prestained or painted shingles against each other and spacing unfinished shingles or shakes ⅛ inch apart. Offset vertical joints between shingles at least 1½ inches from the joints in the undercourse.

3 Installing successive courses. Stretch a string between the next set of story pole corner markings. Then tack a length of the rabbeted siding called shiplap over the existing course, aligning the upper lip with the string. Butt undercourse shingles to the top edge of the shiplap. Butt outer course shingles or shakes to the lip of the shiplap *(inset)*. Remove the shiplap, move the string to the level of the next set of markings and repeat the process. When installing the soffit course and finishing the corners, you should use the techniques for single coursing *(pages 51-52, Steps 5-8)*.

OUTER
COURSE

UNDER-
COURSE

SHIPLAP

Sheathing a House with a Hard Stucco Shell

Stucco—an exterior coating of some hard material—has been used since prehistoric man first daubed mud on his reed huts. More than 2,000 years ago, the Roman architect Vitruvius complained that stucco workmen were becoming slipshod, no longer willing to apply the traditional seven coats made necessary by the thin, brittle lime mixture used then. Today, stucco generally consists of three coats. It still contains some lime, but in composition with sand and portland cement; the material is called mortar, or simply mud, and differs from familiar bricklaying mortar only in its slightly lower lime content.

Stucco has remained popular over the millennia because of its durability and versatility; it makes possible a great variety of decorative surface finishes. But it does require more care and patience to put on than other sidings.

The first, or "scratch," coat is usually ¼ to ⅜ inch thick, and is scored or scratched so that the second coat will bond to it. A ⅜-inch second coat (sometimes called the brown coat because it used to be made with brown sand) requires the most care in application. Thin guide strips of mortar called screeds are first stuck to the wall to keep the brown coat uniform. Mortar is then applied evenly between the screeds. This second coat must be as perfect as possible, because the third, finish coat—a cosmetic layer generally only about ⅛ inch thick—is too thin to conceal imperfections.

To provide a strong base for the first coat on wood walls, waterproof building paper must be nailed over old siding, then covered with heavily galvanized metal lath—tightly stretched to prevent sagging. If the old siding is badly rotted, it must be pried off (page 26), and the building paper and lath applied over the sheathing underneath. Special lathing pieces, called beads, reinforce joints. One type serves for edges and for top and bottom screeds. Another, called expansion bead, is used to prevent cracking where unlike base materials meet, or between parts of a wall supported by separate foundations, such as a main house and a later addition.

Stucco can be applied without lath or scratch coat over masonry or old stucco; the old brick, stucco or concrete block serves directly as a base, and the application is begun with the second coat (pages 58-59, Steps 1-4). To prepare such a wall, scrape off any loose material with a wire brush, scrub it and let it dry. Then, using a paint brush or roller, apply a water-based emulsion, commercially available as "masonry bonding agent." After about 45 minutes it should have a glossy appearance. Touch up any dull spots with a second application and allow the material to dry overnight. An old masonry or stucco wall that cannot be thoroughly cleaned, or that is soft and crumbling, should first be covered with metal lath and then stuccoed in three coats.

Stuccoing calls mainly for common tools; additional specialized ones are widely available at hardware stores or can be rented. You will need a rectangular trowel for most of the job and a small pointing trowel for tight spots. A small mortarboard, called a hawk, is used to hold mortar while you work. You also will need a hoe to mix the mortar, a long wood or metal straightedge, a large, steel-tined metal comb called a scarifier, to score the scratch coat, a browning brush or old whisk broom to dash water onto walls, a mortarboard (a plywood scrap will serve) and mixing box.

Stucco contains lime, which can irritate the skin and burn eyes. Wear gloves and long-sleeved, cuffed shirts while applying stucco and put on goggles when pouring or mixing dry materials. Launder all exterior clothing—it is sure to be spattered with mortar—before wearing it again. Keep water and a rag handy to keep the handles of your tools clean; at the end of each day wash all of the tools thoroughly and hose out the mortar box so dried mortar does not make the next day's batch difficult to mix and apply.

Weather affects stucco. Try to work on a humid, overcast day when the temperature is between 50° and 80°. Never stucco in the hot sun or when the temperature is below 40°. Most professionals consider early autumn the best time of the year to apply stucco, and some prefer to work in a light rain.

A Lath Base for the First Coat

1 **Attaching the foundation bead.** Nail lengths of galvanized stop bead, mesh up, to the foundation 4 to 6 inches above ground level. Use six-penny cut nails at 6- to 8-inch intervals for block foundations; use case-hardened masonry nails, similarly spaced, if you are nailing into poured concrete. Similarly attach stop bead, mesh down, at the top of the wall, pushing it snugly against the underside of the soffit.

BUILDING PAPER

STOP BEAD

2 **Attaching bead around windows and doors.** Use tin snips to clip through the mesh and most of the metal of the bead *(inset)* at points corresponding to the corners. Then bend the bead around the window or door with the mesh facing out. Nail with galvanized roofing nails.

WINDOW CASING

3 **Applying the lath.** Nail sheets of lath to studs horizontally, using furring nails if available in your area, otherwise 3-inch galvanized roofing nails, every 6 inches. Overlap the sheets at least 2 inches at vertical seams and 1 inch at horizontal seams. Tie overlapping seams every 8 inches with 8-inch lengths of 16-gauge wire; tuck protruding wire into the mesh. Butt sheets at corners and against window and foundation bead overlapping the mesh and metal.

LATH

1″ 2″

4 **Attaching the corner bead.** Nail a length of corner bead to the top of the corner so that the nose, or outside corner, of the bead is 1 inch away from the house corner. With a carpenter's level, plumb the bead and attach it down the wall with roofing nails every 6 inches on both sides of the corner. If you use more than one length of corner bead, butt the second piece to the bottom of the first, making sure the nose continues in a plumb, unbroken line.

NOSE

Working with Stucco

Stucco can be purchased in premixed, dry form, but it is more economical to prepare your own. The recipe below makes about 2 cubic feet—enough to cover about 24 square feet—which is all an amateur is likely to use at one time.

The quantities are listed both by weight—sand comes in 50- and 60-pound bags, lime in 50-pound bags and cement in 94-pound bags—and by volume, to simplify measuring with a pail.

Never add additional water to stucco once it has been mixed. If it starts to harden as you are working, retemper it by chopping and mixing it with a hoe.

A Basic Recipe

Building Sand (200 pounds)	15 U.S. gallons
Portland Cement (47 pounds)	3¾ U.S. gallons
Lime (12 pounds)	2½ U.S. gallons
Water	6 U.S. gallons

How to Use a Hawk and Trowel

1 Cutting a hawkful. The techniques of getting fluid mortar onto a wall are the same for all coats and proceed in steps from mortar box to mortarboard to hawk to trowel to wall. After scooping a bucketful of mortar onto the mortarboard, separate about ½ gallon from the pile with the trowel. Then holding the hawk almost perpendicular to the mortarboard, you should move the hawk and trowel toward each other, pushing the mortar with the trowel. As the hawk and trowel meet, tilt both upward slightly, pushing the mortar onto the center of the hawk.

Mixing the materials. Spread the dry ingredients over half the floor of the mortar box. Stand at the empty end of the box and chop the material with a hoe, pulling it toward you until you have shifted it from one end of the box to the other. Repeat two or three times until the mixture is uniform in color. Pour 5 gallons of water into the empty end of the box and chop the dry mixture into it a little bit at a time. Add the remaining water only as needed. After about 10 minutes of chopping and pulling, the mortar should be plastic and uniform in texture. If you pick up a handful and squeeze it, it should leave a fairly heavy residue on your palm, but it will hold together.

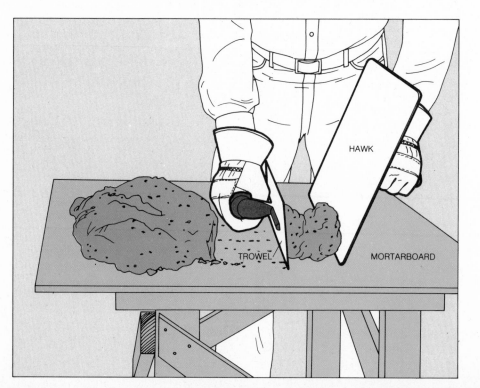

2 Cutting mortar off the hawk. Reach across the hawk and cut into the mortar with the edge of the trowel *(below)*. In one smooth motion, tilt the hawk toward you and away from the wall, at the same time, lifting and twisting the trowel up and away from you. At the end of the motion *(below, right)*, the hawk should be facing you, al-most perpendicular to the ground; the trowel should be blade up with the mortar on the top of it. After each trowelful is lifted, bring the hawk horizontal again and turn it one quarter of a turn so that the next trowelful of mortar is lifted from the far side of the hawk and the diminishing pile of mortar stays centered.

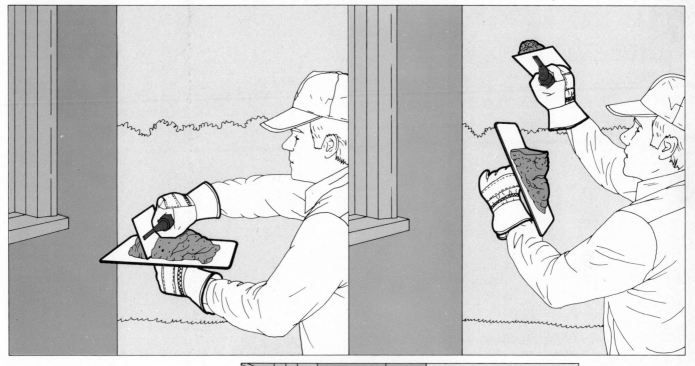

3 Applying mortar. Tilt the trowel back toward the wall, placing the bottom edge against the sur-face at a slight angle. With a smooth, continuous motion, push upward, forcing the mortar into place. Apply overlapping vertical sweeps from right to left if you are right handed, left to right if you are left handed. Finally, with the trowel al-most parallel to the wall, gently smooth the ridges left by the overlapping strokes.

Applying the First Coat

A scratch coat for wood walls. If you are putting stucco over wood, nail lath and beads to the wall (*pages 54-55*) and apply a ⅜- to ½-inch layer of mortar—most of it behind the lath—using the techniques described on pages 56-57. The mortar should cover the lath so thinly that a faint impression of the pattern of the lath shows through. While the mortar is still damp, score the scratch coat horizontally with a steel-tined scarifier. Press hard enough to contact the lath, but not enough to expose it. Allow the scratch coat to dry at least 4 to 6 hours.

If you are putting stucco over a sound masonry wall, no scratch coat is needed; simply apply screeds and a brown coat directly to the wall as described below.

Applying the Second Coat

1 Fixing the screeds. Place strings horizontally to establish a plane for the brown coat, which goes over the scratch coat on a wood wall but directly over a masonry wall. Pass the strings across the wall, stretching them taut around the corner beads, which will hold them the proper distance from the wall. Secure the strings to nails around the corners. The first string should be 5 feet below the roof line, the second—and third if needed—spaced 5 feet away down the wall. Space nails every 5 feet along the strings, driving them only far enough to bring their heads even with the strings. If a window or door interrupts the spacing, place the nails a foot from it and continue the pattern beyond it.

Remove the strings, leaving the nails to guide you later, and shake water onto the wall between the nails with a browning brush. Then trowel on thin vertical strips, or screeds, of mortar, bringing them flush with the nailheads.

STRING

CORNER BEAD

2 Leveling the screeds. Hold a straightedge against the wet screeds with the ends pressed against the beads or nailheads and wiggle it back and forth slightly to cut off any excess mortar. If there are any gaps between the straightedge and the screeds, fill them in with more mortar. Frequently scrape the straightedge clean with a trowel, flipping the excess mortar back onto the mortarboard. Let the screeds dry for 24 hours before applying the brown coat.

3 Applying the brown coat. After dampening the screeds and scratch coat (or masonry wall), trowel mortar onto the top of the wall flush with the beads, feathering it into the screeds. If you have trouble blending the mortar and screeds, shake water onto the screeds with a browning brush. Check the depth of the new mortar frequently as you apply it, running the straightedge over the screeds and beads. Wiggle it to cut off excess mortar; add mortar to fill any gaps.

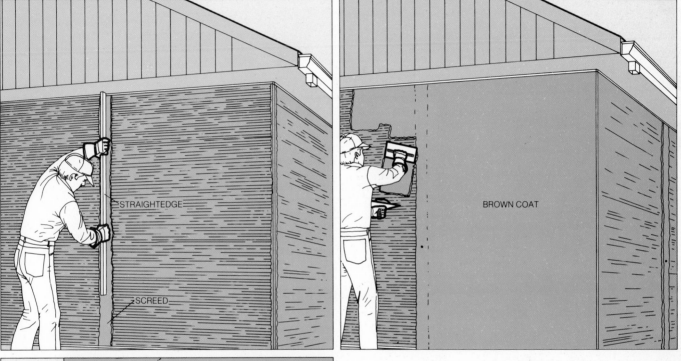

STRAIGHTEDGE

SCREED

BROWN COAT

4 Floating the brown coat. About an hour after the brown coat has been applied, smooth its surface with a special implement called a wood float. Work the float in light, circular motions to leave an even, slightly sandy-feeling surface.
Spray the brown coat lightly with a garden hose every 12 hours for 48 hours and then let it dry for another five days before applying the finish coat as explained overleaf.

A Variety of Finish Coats

When you apply base coats of stucco, there are certain rules you must follow to ensure a siding that is sound, plumb and weathertight. But when you apply the thin, cosmetic finish coat—a layer that is purely decorative—you can make your own rules. If you use the time-honored troweling and texturing techniques, you can achieve any of a number of traditional finishes. One of the most popular, the smooth float finish, is applied exactly like the brown coat *(pages 58-59)*, but is smoothed with a special sponge- or cork-covered float. For more daring textures, you can slap mortar on with a brush, embed pebbles in the wall or finger-paint the texture on with gloved hands.

Many traditional finish coats are associated with a particular style of architecture. The travertine, which imitates the heavily veined texture of Italian travertine stone, is found on houses of Mediterranean design; the rough, rustic English cottage texture is often used with half-timbered, Tudor style architecture.

Most finish coats are troweled on like base coats, and then one of a variety of implements, from whisk brooms to wood blocks, may be used to texture the surface. Stippled and scraped textures are easiest; smooth finishes and heavily textured finishes like the spatter dash or pebble dash *(opposite, top)* are trickier.

Regular stucco mortar is used for most finish coats. Special finishes, though, may require a different formulation. For instance lime, which is added to regular mortar to make it easy to work with a trowel, is left out of the mix for the pebble-dash finish, one that is thrown rather than troweled onto the wall. For a white finish, use white portland cement and white sand instead of the usual building sand and gray cement. For a colored finish, mix the mortar with commercially available colored cement.

The finish coat must be cured with special care. No matter which finish you use, do not disturb it for 24 hours after applying it. Then lightly mist it with a garden hose 3 or 4 times over the next 24 hours to keep it barely moist. If the weather is windy or stormy, protect the walls with plastic sheets.

English cottage. After allowing a ⅛-inch mortar coat to harden about ½ hour, dab on chunks of mortar with a roundnosed trowel. Hold the trowel at an angle and twist it to form irregular ridges. You can control the rough, rustic look of this texture by varying the amount of mortar on the trowel and the pressure of the strokes.

Modern American. Trowel on a ¼-inch coat of mortar, allow it to set until the surface moisture disappears and scrape the surface with a wood block. Hold the block at right angles to the wall and scrape with straight, upward strokes. The surface will take on a rough, torn look and should be free of smooth spots.

Spatter dash. Holding a browning brush or a whisk broom full of mortar in one hand and a stick in the other, snap the edge of the brush against the stick to dash the mortar in a thin even spray to cover the wall evenly. After about one hour, dash on a second coat for uniform depth and texture. Drape windows and doors with heavy paper or burlap before dashing mortar around them. For the similar pebble-dash finish, make a soft, runny mortar of half portland cement, half ⅛- to ½-inch pebbles and water—use no sand or lime. Dip the material out of the bucket with a garden trowel and dash it at the wall with a sweeping, sidearm motion.

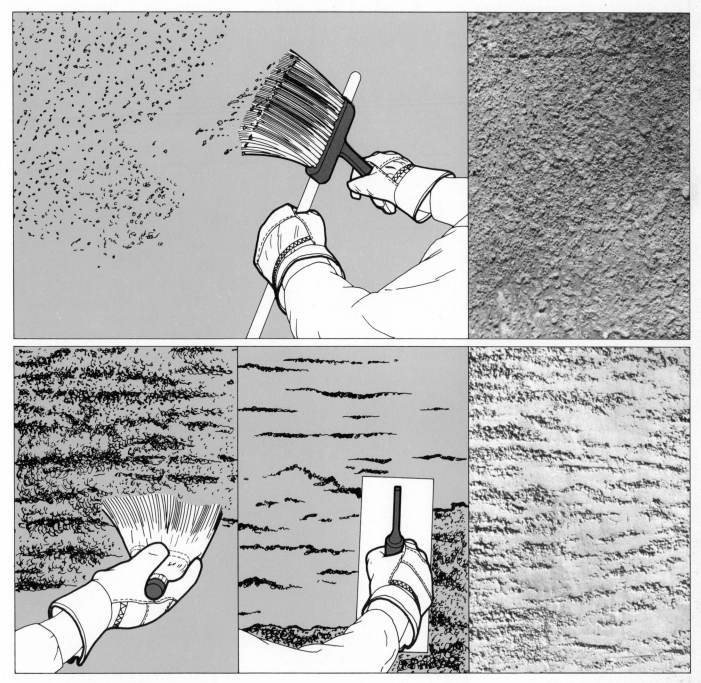

Travertine. Apply a ¾-inch-thick coat of mortar and trowel it fairly smooth, then jab it deeply with a stiff brush to produce a rough, stippled texture (*above*). As the mortar sets, within about half an hour, lightly smooth out the higher portions of the texture with a clean trowel (*above, right*), moving it in even, horizontal strokes.

Asbestos Shingles

Asbestos shingles are fire- and rot-proof, and they do not need painting. Though they are brittle, chipped or broken ones are easy to replace because the nails are exposed. Simply break the damaged shingle free of the nails *(page 26)*, remove them, slide in another shingle and nail it down.

The most common asbestos shingles are $3/16$ inch thick, 12 inches wide and 24 inches long, though some styles are larger and thicker. They come in lots to cover 100 square feet, complete with asphalt-impregnated backer strips for joints and tight-gripping annular nails that match the color of the shingles.

Because asbestos shingles are brittle, they cannot be applied directly over old clapboard or wood shingles as most other siding can—nailing to the projecting angles of lapped materials would leave unsupported the tops and bottoms of the untapered asbestos shingles, and they could easily break. They should be nailed to smooth wood or fiberboard sheathing (use special nails available from building-supply dealers), or in the case of stucco walls, to furring strips. Except in the case of stucco, it is generally advisable to remove the old siding and attach the asbestos shingles to the sheathing underneath.

There are three ways to finish corners when installing asbestos shingles. You can use wood corner boards *(page 28)*, individual metal corners similar to those for clapboards *(page 40)* or special metal corner beads. The latter method, illustrated on these pages, is simplest.

When trimming, shaping or making holes in asbestos shingles, always use a shingle cutter *(above, right)*. Available from tool-rental stores, a shingle cutter creates almost no asbestos dust, which can cause serious lung damage if inhaled. For extra security, wear a respirator when using the cutter.

Using a shingle cutter. To trim a shingle, set the cutter blades to the shingle thickness with the adjustment knob and use the cutter as you would a paper cutter.

To shape the edge of a shingle to fit around an obstruction such as a window, nibble off small bites (to avoid breaking the shingle) with the hook-shaped nipper on top of the cutter.

For making nail holes, the cutter has one punch under the handle and another near the pivot, which pokes a hole 1 inch from the shingle edge. To perforate a shingle, align it with one of the punches and depress the handle.

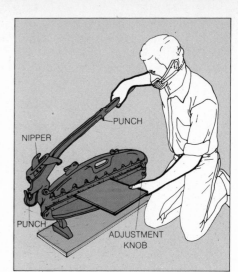

Installing the Shingles

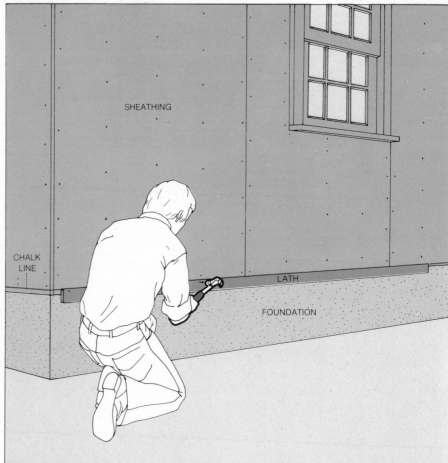

1 Starter strip. With fourpenny nails, attach a wood-lath starter strip, $3/8$ inch by $1\frac{3}{4}$ inches, for the first course of shingles $3/4$ inch above the joint between the foundation and the sheathing, using a chalk line to align it. Stop the starter strip short of corners to leave space for metal corner bead or wood corner boards. If the surface to be shingled is wood or stucco, cover the wall with a layer of 15-pound roofing felt so that it protects the lath. (Felt is unnecessary over fiberboard sheathing because it is waterproof.) To be sure that the felt is weathertight, overlap the ends and sides of successive felt strips by 4 inches except at corners, where the overlap should be 6 inches. Nail on the metal corner bead (or wood corner boards as shown on page 28).

2 Starting the first course. Slip a shingle into the corner bead so that it overlaps the bottom of the starter strip by ¼ inch. A chalk line 11¾ inches above the bottom of the starter strip is a handy guide for positioning the shingle. Working from the corner, drive nails into the first two holes in the shingle bottom. Drive nails snug but not tight, to avoid breaking the shingle.

3 Flashing the joint. Slip a backer strip halfway under the end of the shingle, then hammer the third nail. Finish the course in this manner, always positioning the backer strip before hammering the third nail. Trim the last shingle to fit the leftover space at the end of the course and punch it with a nail hole 1 inch from the trimmed end and the bottom.

CHALK LINE

4 Installing succeeding courses. Cut a shingle in half to start the second course and punch a nail hole in the bottom corner. Push nails partway through the holes, then position the half shingle on the wall by resting the protruding nail tips on the edge of the shingle in the course below. Slip a backer strip under the shingle edge so that it overlaps the course below about ¾ inch, then hammer in the nails. Finish the course with whole shingles, then alternate between whole and half shingles to start the remaining courses.

5 Working around windows and doors. Use the nipper of a shingle cutter to fit shingles to corners of windows and doors. Below windows, measure from the top of the preceding course of shingles to the window sill and trim shingles one inch taller. Punch nail holes in the top edge opposite the ones in the bottom edge and nail the shingles to the wall through both sets of holes. At the sides of windows and doors, trim shingles to preserve the staggered pattern of joints. On top of windows and doors, shim the shingles with scraps to match the slant of adjacent ones, then nail the shingles through the top.

For the top course, fit shingles below the soffit or fascia board as you did below the windows.

Breaking with Tradition

Traditionally, builders have been concerned with tradition. Ancient Egyptians carved stone to look like wood, imitating primitive reed and mud construction details. Many centuries later, George Washington covered Mount Vernon with wood carved to look like stone. Today roof and siding manufacturers respond to popular yearnings for tradition by taking modern materials and fashioning them in conventional forms. Aluminum or vinyl siding is made to look like wood clapboard, asphalt like wood shingles; hardboard panels are textured like stucco, and plastic and fiberglass are molded to look like stone or brick.

The force of tradition is equally powerful in conventional associations between materials and architecture. Whether it is an old natural substance or a new imitation, the material in which a house is clothed often dictates—or is dictated by—the style in which it is built. Clapboard is equated with Cape Cod style, brick with Federal or Georgian and stone with Country Manor.

But today several factors are at work to free materials from tradition. One is the flow of new ideas and materials from industrial and commercial builders into the normally conservative housing market. Offered a wealth of new choices, architects are free to sample freely from the industrial marketplace. They can, for example, build a house of structural steel—a house whose walls, like those of skyscrapers, are thin glass membranes serving a protective rather than a structural function. They can fashion an entire house of molded plastic, forming a thin weatherproof shell that, despite its unconventional appearance, offers a home-owner clear practical advantages of easy maintenance or superior insulation. They can even build homes with walls and roofs that function as effective solar heating and cooling systems.

Inspired by new materials and new techniques, both architects and home-owners are also taking a fresh look at old-fashioned materials and using them in imaginative and unexpected ways. The age-old surfaces of sod and stone turn out to look better than ever on structures built today. Latticework, seemingly more at home on the huge verandas of fashionable 19th Century hotels, can serve a delightful decorative function when it is applied to a modern home. A scalloped pattern in decorative wood shingles, commonly associated with beach cottages, becomes a surprisingly appropriate finish for a stark contemporary structure.

Finally, builders have begun to capitalize on the unexpected beauty of modern construction work horses such as plywood, aluminum and concrete block. In some cases, paradoxically, they are only rediscovering old beauties. Aluminum was once considered so precious and lovely that in the 1855 Paris Exposition, bars of it were displayed next to the crown jewels; soon afterward, Emperor Napoleon III used aluminum flatware to serve his favored dinner guests. Fine laminated woods, the precursor of modern plywood, have been used for centuries in the manufacture of elegant furniture. It has only been in the past century, as these materials became commonplace, that they acquired a humdrum, utilitarian reputation. Today, in creative hands, they are once again being treated as versatile ornamental resources.

A lacy illusion. For a Virginian who wanted "the romance of a porch," an architect created a house-within-a-house, using an audacious mix of old and new materials. An old-fashioned latticework shell, made of redwood strips tacked to wood framing, surrounds a multitiered residence clad in modern painted plywood. Flat interior roofs serve as decks, and the entire structure is topped by a vast shed roof sheathed in translucent plastic.

A

Unexpected Beauty
in Common Materials

Concrete, asphalt, aluminum and plywood are the staples of modern residential construction, exploited for their strength, durability and moderate price. For the most part, their esthetic qualities are ignored or concealed. Baked enamel covers aluminum; concrete block and plywood hide behind traditional veneers of brick or wood.

Yet in the three contemporary homes shown on these pages, each material displays striking decorative power. And each, despite the popular notion that modern materials intrude upon natural forms, blends into its surroundings.

Below, anodized aluminum, undisguisedly metallic, sheathes the side walls of a lean structure of glass and industrial steel opening onto a leafy, wooded site. Unadorned concrete-block walls *(opposite, top)* set in lush, tropical vegetation take on the solidity and repose of massive boulders. And homely plywood siding under an asphalt-shingle roof *(opposite, bottom)* nestles peacefully into a grassy, wind-swept hillside.

Open house. Glass sheathes most of the house; the only opaque walls consist of unpainted aluminum siding, anodized to prevent corrosion, and 16 domed plastic skylights in the roof open the house to the sky. A structural steel skeleton and exposed steel joists set the tone of spare, open elegance that characterizes this home in Toronto.

Tropical building blocks. The practical virtues of concrete-block construction are exploited in this Miami residence. Exposed both inside and out, the concrete is impervious to the rot that plagues homeowners in humid areas. To help maintain a comfortable climate indoors, the cells inside the blocks are filled with insulation.

Plain geometry. All straight lines and angles, the clean-cut structure of this house in Northern California is enhanced by the simple, practical materials in which it is clothed. Finished in long continuous strips, plywood makes an ideal material for the seamed batten siding, and precisely fitted asphalt shingles give the sweeping roof a clean, workmanlike appearance.

Materials of the Future: Saving Energy and Time

The materials that cover houses on these pages serve simultaneously as structural elements, insulation and siding. Both the gaily striped little house at right and the imposing home below are made of plastic, quickly assembled and resistant to the elements. As a bonus, the house at right requires little maintenance beyond an occasional rinse with a hose. The two New Mexico homes on the opposite page have walls that function as ingenious systems for tapping solar energy and controlling summer heat.

A foam home. The seamless walls and roof of this Connecticut house are made of a dense, insulating industrial foam sprayed over a netting of dacron rope and fiberglass fabric. Within seconds, the foam dries and the structure is sprayed with a resinous, stucco-like plaster.

An instant house. A British experiment in low-cost mass housing, this building comes from a mold. Poured in a factory, it is delivered intact and simply tethered to its foundation. The polyester material is light yet strong and can be colored in various shades and patterns.

Solar walls. Water-filled metal drums behind the glass walls of this house absorb the sun's heat until nightfall, when huge doors rise to insulate the drums from the outside, and their heat radiates into the rooms. In summer, conversely, night-chilled water offers daytime cooling.

A wall of beads. Watched by a curious cat (*above*), tiny styrofoam beads are sucked down and out of a cavity between two panels of glass. On a summer day, the beads are blown into the cavity (*left*) to insulate the house from the heat of the sun; in the evening they are drained to let the house cool off. In winter, the walls are filled at night to retain daytime heat.

One Foot in the Past, One on Solid Modern Foundations

Space-age dugout. Dug into a sand dune on the Florida coast, two small apartments peer out of their burrow at the ocean. The gently curved walls, their shape and strength calculated by computer, are sprayed concrete, and the grass and vine ''siding'' control erosion and preserve the site's natural character.

Earth, shingles and stone are among the most traditional of all building materials, and each is commonly associated with a traditional—even archaic—building style. Yet these time-honored materials have unexpected possibilities as vehicles for contemporary architectural effects.

The striking building at the top of the opposite page effectively dispels any association that shingles may have with quaint Victorian or seaside architecture. The stone house *(opposite, bottom)* has little in common with another architectural cliché, the rustic stone cottage. Computer technology helped to create the earthen structure below, resurrecting the prairie sod house without its leaks and structural failures.

A house of curves and cylinders. Individually applied round-cut cedar shingles wind around a residence in Redondo, Washington, their wood tones warming and softening the stark form of the structure. The small curves of the shingles complement the massive cylinders of the walls with an interplay of light and shadow.

Tied to the land. A contemporary residence in Sardinia's Costa Smeralda resort area owes much of its distinction to walls of rough native granite. Perfectly suited to the untamed, boulder-strewn land that surrounds it, the house ignores the whims of fashion to set its own timeless standards of beauty and strength.

G

3 Roofs that Resist the Elements

A roofing sampler. Though they all do the same job of keeping off the wind and the weather, these materials vary strikingly in appearance. Arranged on a sheet of asphalt roll roofing, they are, clockwise from the top: ceramic tiles, a galvanized steel panel, slates, a variety of roofing nails and asphalt shingles. Absent are wood shakes; they appear on the cover of this book.

Over its lifetime it is drenched by more than a million gallons of water weighing almost 5,000 tons, baked by 50,000 hours of direct sunlight, stretched and squeezed by changes of temperature totaling a half a million degrees and swept by two million miles of passing wind. Only then—after 20 years—is an asphalt-shingle roof ready for replacement. Roofs of slate and tile do even better; they can shrug off nature a century or more.

Putting up roofs to withstand such punishment used to be the exclusive province of a skilled craftsman, who pieced each roof together from small, irregular units, locally made and individually fitted. No longer. Over the years the job has been made easier for both professional and amateur by standardized roofing materials designed for easy installation.

Asphalt shingles, in a wide assortment of styles, colors and weights, are appropriate to an increasing variety of roofing situations. The shingles are easily applied and come with dabs of adhesive that automatically seal the roof against high winds.

Metal roofing materials are constantly improved. Traditionally, they have consisted of galvanized steel, laid on in corrugated sheets with large overlaps or of flat panels, laboriously soldered together. Both types are giving way to nonrusting aluminum panels with interlocking joints that eliminate soldering, minimize overlap and prevent water from siphoning through the seam sideways.

Slates now come in standard sizes with prepunched nail holes. Imitation slates made of molded asbestos offer the hand-cut look of slate at a lower cost.

Tiles are manufactured in classic patterns that fit together like building blocks.

Even the manufacture of wood shakes, once a typical cottage industry, has been revolutionized by the machine. The classic shake, a thin slab split from a section of a cedar log, was formerly made with a wide-edge hatchet called a frow. Sections are now split on a hydraulically lifted table that bumps the wood against a knife, and the resulting slab is then run through a band saw to make two tapered pieces. These modern shakes go onto a roof more easily and tightly, but retain their hand-split ancestors' rustic look, thick butt and the merits that result from having been split: natural rainwater channels and (because the wood cell walls are left intact) superior resistance to absorption of water.

Some roofing installation—such as built-up roofing *(page 84)* and soldered metal roofing *(page 98),* both common on flat roofs—remain beyond the reach of most amateurs. But even they can be repaired by the homeowner.

Choosing the Right Roofing, Preparing the Roof

The choice of a roofing material, like the choice of a siding, involves such factors as appearance, cost, durability and ease of installation. But in roofing, an added factor—the slope of the roof you plan to cover—limits the choice of materials.

The more nearly level the slope, the slower the runoff of water from the roof, and a slow runoff calls for an especially waterproof covering. A completely flat roof, for example, must be covered by soldered metal (pages 98-99) or by built-up roofing (pages 84-85)—and because these materials need special installation skills and tools, a flat roof should be laid by a professional. But a roof with a gentle slope can be covered by a homeowner, using roll roofing (pages 82-83).

On steeper slopes, the range of practical materials increases. Asphalt shingles, for example, are widely used on roofs with moderate to steep slopes because they are inexpensive, easy to install and available in a wide variety of colors. To enhance the value of your house, you may prefer to reroof with such traditional materials as slate, ceramic tile, or wood shingles or shakes. All are attractive and durable—but they are also expensive, hard to install and may need special roof preparations (pages 68-71): slate and tile are particularly heavy and require sturdy roof supports, while wood shingles must be laid over open planks for ventilation.

The chart on the next page compares the factors of cost and convenience in widely available roofing materials; discussions of each material and its method of installation make up the bulk of this chapter. The drawings on this page show how to measure the slope and area of a roof—information you need in order to choose a roofing material and to estimate the amount to buy. Most materials are sold in units called squares, each capable of covering 100 square feet; the only exception is roll roofing, which comes in rolls of varying lengths and weights.

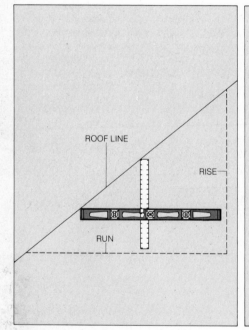

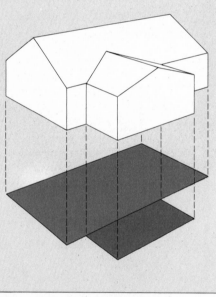

Slope	Conversion factor
1 to 3 in 12	1.03
4 in 12	1.06
5 in 12	1.09
6 in 12	1.12
7 in 12	1.16
8 in 12	1.20
9 in 12	1.25
10 in 12	1.30
11 in 12	1.36
12 in 12	1.45

Measuring the slope of a roof. A roof's slope is the rate at which it rises, expressed as inches of vertical rise for each foot of horizontal run; for example, a roof that rises 3 inches vertically while covering 12 inches of the house beneath it is said to have a slope of 3 in 12. To measure slope from outside a house, mark off 12 inches on a level and set the end of the level at the rake of the roof. With the level itself aligned horizontally under the rake, use a ruler to measure the vertical distance between the 12-inch mark and the rake (in the example above, a 10-inch reading on the ruler would indicate a slope of 10 in 12). To measure a slope from inside a house, follow the same method but measure along a rafter rather than the rake.

Measuring the area of a roof. Make a plan of the ground space covered by each surface of a roof, including overhangs at gables and eaves. Break the ground spaces down into rectangles and calculate the area of each rectangle. In the example above, the house occupies two rectangles, covered by roofs of two different slopes.

To convert flat ground areas into slanted-roof areas, use the table above, right. Find the figure in the right-hand column that corresponds to the slope of the roof, and multiply this figure by the ground area under the roof to get the area of the roof. To estimate the amount of material you need to cover roofs of different slopes, take the total of the roof areas, add 10 per cent to

this total to allow for double layers of covering along ridges, eaves and hips, and round the new total to the next higher square—that is, the next 100 square feet—of material.

In the house shown here, one of the rectangles measures 1,600 square feet, covered by a roof with a slope of 10 in 12; the other measures 400 square feet, covered by a roof with a slope of 5 in 12. The area of the first roof is 1,600 by 1.30, or 2,080 square feet; that of the second is 400 by 1.09, or 436 square feet; and the total roof area is 2,516 square feet. An additional 10 per cent allowance makes the total 2,768 square feet, which should be rounded to 2,800 square feet, or 28 squares.

A Guide to Roofing Materials

Type	Cost	Durability (years)	Minimum slope	Advantages	Limitations
asphalt shingles	inexpensive	12-25	2 in 12	easy installation; available in a variety of weights and colors; requires little maintenance; easy to repair	poor fire resistance
roll roofing	inexpensive	10-15	1 in 12	easy installation and maintenance	poor fire resistance; drab appearance
built-up roofing	moderate	10-20	0 in 12	most waterproof of all roofing	poor fire resistance; must be installed professionally; leaks difficult to locate
wood shingles and shakes	moderate to expensive	15-30 (shingles) 25-75 (shakes)	3 in 12 (steeper in humid climates)	easy installation; attractive rustic appearance; natural insulator	highly flammable unless specially treated; shingles must be laid over open planks (page 69)
slate	expensive	50-100	4 in 12	attractive traditional appearance; fire resistant	heavy, brittle; requires sturdy roof support; long and delicate installation may require special tools; needs regular replacement of damaged pieces; difficult to repair
ceramic tiles	expensive	50-100	4 in 12	attractive traditional appearance; fire resistant	heavy, brittle; requires sturdy roof support; time-consuming installation requires special tools; availability of replacement pieces unreliable; difficult to repair
metal panels	moderate	25-50	2 in 12	easy installation and patching; can be painted any color; fire resistant	subject to damage from wind, trees, any contact

Comparing roofing materials. In this chart, "Cost" refers to the relative cost of materials alone; it does not include the cost of labor. In most cases, the cost of professional installation is higher for the traditional roofing materials—slate, tile and wood shingles and shakes—than for the newer materials, which are designed to be installed more quickly. The minimum slope is the slope at which a specific material begins to provide adequate protection against water. All the materials listed can be applied to surfaces steeper than the minimum, but as slopes increase, such considerations as appearance and durability become more important. Roll roofing, for example, provides adequate covering for steep as well as gentle slopes, but its plainness and poor durability make it an unlikely choice for any but the most gradual slopes.

"Durability" is a rough measure of how long a roof will last with proper maintenance; the figures given apply to temperate climates. The columns listing advantages and limitations for each material concentrate on installation and maintenance.

Making a Sound Roof Deck

Often a new roof can be laid directly over the old one, with no more preparation than the removal of material at the ridge and hips *(below)*, and the installation of new flashing *(pages 16-21)*. But if there is already more than a single layer of roofing in place—as is likely on any house more than 20 years old—you must pry off all old material down to the wood sheathing covering the rafters. And sometimes sheathing *(opposite)* or rafters *(pages 72-75)* may also need repair. Both may rot and the roof may take on a wavy, uneven appearance, or even sag.

This understructure, called the roof deck, should be checked before new roofing is laid—if necessary, by removing parts of ceilings. From an unfinished attic, stick an awl into rafters and sheathing from top to bottom on each side and along the ridge beam. If they are rotted, the awl will penetrate the wood easily.

At the same time, look for stains that indicate recent wetting. If moisture stains indicate a leak from above, the new roofing will solve the problem. But if moisture does not appear to be from a leak, additional venting may be needed.

In some instances you may have to replace the roof deck to apply new roofing, even if the deck is in good condition. All roofing except wood shingles and shakes requires solid sheathing, preferably of plywood *(opposite, center)*. Shingles and shakes are generally laid over an open roof deck made of 1-inch-thick boards spaced 2 to 4 inches apart *(pages 86-90)*.

If you have an open deck you can re-roof with wood, using the existing boards, provided you keep the same distance between courses. If the spacing will change, move the slats or replace them *(opposite, bottom)*. If you are re-roofing with any other material, provide a solid deck by replacing the open deck with plywood or by nailing plywood over the slats. Caution: remove and replace sheathing one side at a time only; removing all of the sheathing at once may allow the roof to shift, or even collapse.

Once the deck is prepared, protect its perimeter with drip edge, metal preformed to direct runoff away from fascia boards at eaves and rakes.

Underlayment—which is generally a heavy, asphalt-impregnated paper called roofing felt—is used under the roof except for wood or roll roofing. It adds protection to the roof and shields it from rain while the new roofing is installed.

In any climate, low-sloped roofs—those with pitches of less than 4 inches per foot—call for a double layer of underlayment *(page 71)*. For steeper pitches in climates generally free of ice, apply one layer of felt. If there is occasional icing, add a layer of roll roofing at the eaves *(page 70)*. Where icing is severe, you will need a double layer of felt over the entire deck. Use 15-pound felt underlayment—it weighs 15 pounds per 100 square feet—for most roofs; 30-pound felt is recommended for slate roofs and 45-pound felt for tile roofs.

Stripping the Roof

Removing old roofing. Work a shovel or spade under the roofing—all types are handled similarly—at the ridges until it catches against the roofing nails. Pry up the nails. Then the ridge or cap must be removed before new roofing can be applied, whether the rest of the old material is taken off or not.

If there is more than one layer of old roofing in place, remove it all, working from the top of the roof to the bottom to take away successive courses. If you remove all old roofing also rip off the old felt and remove old flashings.

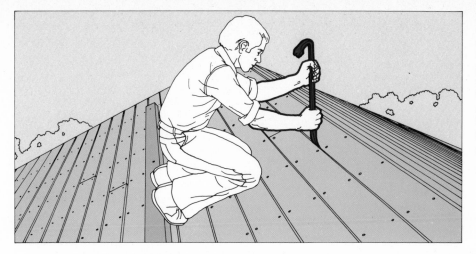

Prying off sheathing. If sheathing must be removed, work the straight end of a pry bar into a crack between boards or sheets of plywood and raise the sheathing up just far enough to begin easing the nails free. Hit the board back down with a hammer. The popped-up nailheads can then be pulled with the claw of the pry bar. Alternate the operations, lifting the board and pulling nails until the wood is loose enough to pull free by hand. If rafters must be repaired or replaced, see page 72.

Replacing the Wood Deck

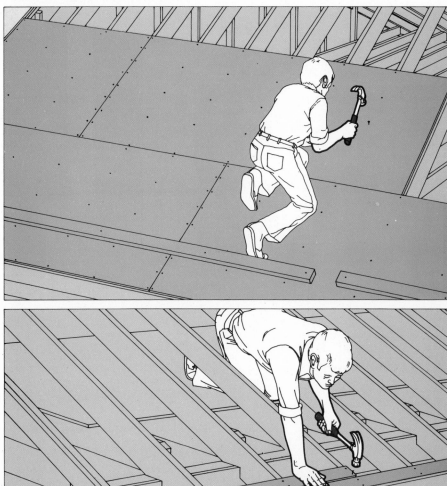

A solid deck of plywood. Starting at a rake lay Type C-D plywood sheets along the eaves with the long edges horizontal. Leave a space between sheets of about 1/16 inch in most areas, but double that in very humid climates. Nail at the edges every 6 inches, and at rafters in between every 12 inches. Stagger joints as you nail other rows up the roof, temporarily nailing 2-by-4s over the sheathing for footholds. At the ridge, trim the sheets even with the center of the ridge beam. Nail patches of sheet metal over large knot holes or sappy spots, and install new flashing *(pages 16-21)*.

An open deck for wood shingles. Using 1-by-3s for 16-inch shingles or 1-by-4s for larger ones, attach boards along the eaves with two nails at each rafter; leave a 1/8-inch gap at joints. Attach a second row so that the distance from its center line to the eaves is equal to the shingle exposure *(page 86)*. Continue to nail boards up the roof, separating their center lines by the exposure distance and staggering the joints. A scrap of wood makes a handy measuring guide. Nail two boards side by side along the ridge. Install new flashing *(pages 16-21)*.

Edging the Deck with Metal

Installing drip edges. Fit preformed metal drip edge snugly against the fascia board at the eaves, and nail it every foot with roofing nails centered in the top surface. After applying underlayment *(below)*, add drip edge to the rakes.

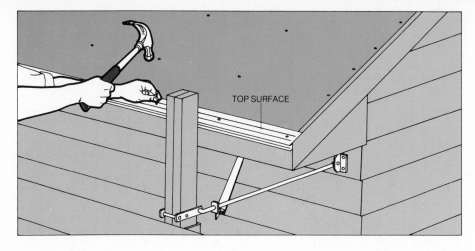

Adding a Layer of Felt

A single layer of felt. For all roofs except those covered in roll roofing or wood, at least a single layer of underlayment is used (additional protection may also be needed as described below). Nail the first strip along the eaves, using roofing nails or special underlayment nails every 2 feet along the bottom and middle of the strip. Overlap subsequent strips 3 inches, trimming the last strip to overlap the ridge by 6 inches. If you reach the end of a roll in the middle of the roof, overlap the new sheet 4 inches and nail every 6 inches along the seam.

Reinforcing the eaves. In areas where icing occurs occasionally, nail a strip of 50-pound roll roofing over the felt along the eaves *(right)*. If the overhang is 2 feet or more, nail a second strip to overlap the first by at least 6 inches, making sure the upper edge of roll roofing extends at least a foot upward past the exterior wall underneath. Seal the seam with roofing cement.

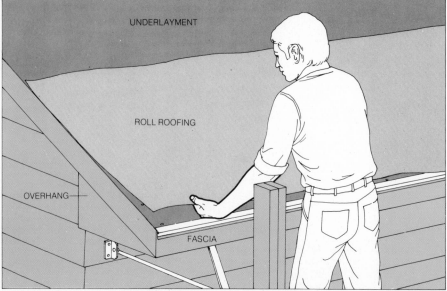

Installing double underlayment. On all roofs with pitches less than 4 inches in 12, and on more steeply pitched roofs in extremely cold areas, use two layers of felt over the whole roof. To install a double layer, nail a starter strip, cut 19 inches wide, along the eaves. Coat it with roofing cement using a comb trowel; apply about 2 gallons for every 100 square feet. Nail a full strip over the starter strip. Then coat the upper 19-inch portion of the second strip with cement. Lap each succeeding strip over the previous one by 19 inches, cementing the upper portion of each until you get 2 feet beyond the wall below the eaves. Then simply nail the overlapping strips without cement.

STARTER COURSE

Marking the Felt with Guidelines

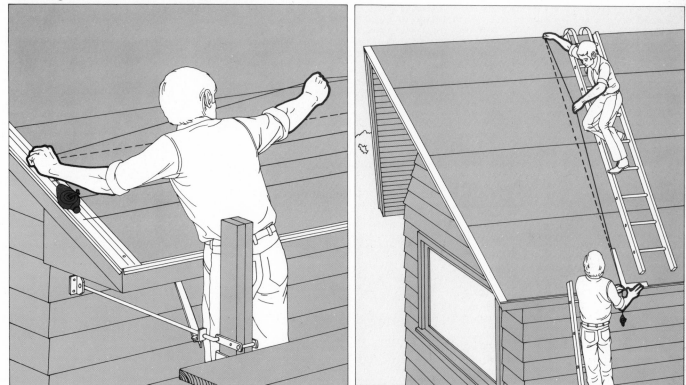

Making horizontal guidelines. For asphalt shingles, snap a chalk line 11½ inches above the eaves and parallel to them. Then snap parallel chalk lines every 5 inches up to the ridge. If one rake is slightly shorter than the other, reduce the space between the lines on the short side so that the top line will be parallel to the ridge. For slate, subtract 1 inch from the nominal length of the slates and make the first chalk line this distance up from the eaves. The distance between subsequent lines should equal the exposure of the slates (*page 92*). Horizontal guidelines are not needed for wood shakes or shingles; tiles call for the special markings described on page 95.

Marking vertical guidelines. To find verticals on roofs having rakes that are not at right angles to the eaves, align one leg of a steel square with the eaves; hold one end of a chalk line at the corner of the square while a helper on the ridge adjusts the other end. When the line parallels the vertical leg of the square, snap it. On roofs with perpendicular rakes and eaves, simply measure in equally from the rakes at top and bottom to position the chalk line. Mark two or three verticals on each slope.

Repairs for Damaged Rafters

Usually rafters will last for the lifetime of a house. But when you inspect a house before reroofing, you may find damage that should be corrected before a new roof is put on. Spots that have been wet but have not rotted can simply be allowed to dry thoroughly and then treated with wood preservative. But advanced rot, insect damage or cracks or bows caused by the drying of rafter lumber over the years call for further action. Small sags can be corrected, but large ones require professional help.

A rafter damaged near the middle can be reinforced with "sisters." These are new, partial rafters nailed to the sides of the old rafter, sandwiching the damage (below). If the damage is near the ridge beam, the sisters must be cut with beveled ends and nailed to the good wood of the old rafter below the damage and to the ridge beam above it. If the damaged area is near the eaves, the ends of the sisters must be cut to fit the top plate of the house wall and nailed to the old rafter above the damaged area. When the damage is extensive a sister must be fitted to both the ridge and the top plate. Such a full-length sister will take over the load of the old rafter, but making the full-length sister requires some tricky cutting and fitting. If the sheathing is to be taken off, however, the old rafter can be removed and used as a template to cut a replacement rafter that can easily be nailed in from above.

You can correct for a bowed rafter without removing the sheathing with a technique similar to that used in installing sister rafters (below).

To make sister rafters, use lumber that is the same nominal size as the old rafters. The boards probably will be slightly smaller than the old ones because lumber sizes have been reduced, but an exact match is unnecessary.

Use only sister rafters, rather than replacement rafters, to correct damage to a truss roof, one built without a ridge board and supported by rafters tied together with a web of 2-by-6 reinforcing members. Removal of any part of the truss structure can seriously weaken the roof. Do not confuse the multiple reinforcements of truss roofs with collar beams—simple horizontal members that tie pairs of rafters together. Collar beams can be removed and replaced as necessary. Damage to the ridge board, on the other hand, is a job for the professional.

Reinforcing a weak rafter. While a helper holds a sister rafter—a board the same width as the existing rafter and cut long enough to extend 2 feet above and below the damage—next to the existing rafter, nail the two together; use nails long enough to penetrate an inch beyond both rafter pieces and clinch the exposed points. Space the nails at 6-inch intervals near the top and bottom edges of the rafters. Then spike a second, identical sister onto the other side, using similarly spaced nails long enough to penetrate into but not through all three boards.

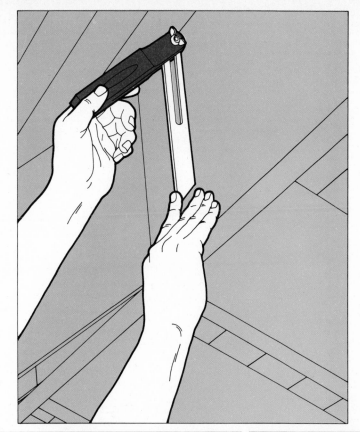

Sister Rafters to Bear the Load

1 **Finding the rafter angle.** To install a full-length sister rafter, first establish the roof angle with a T bevel. Hold it with the blade against the ridge beam and the handle against the underside of the sheathing. Tighten the wing nut to hold the blade in position.

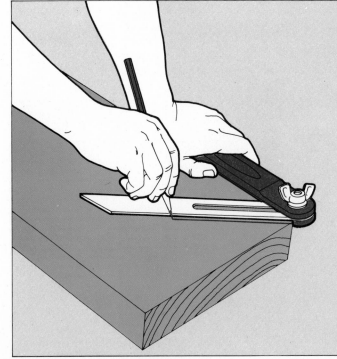

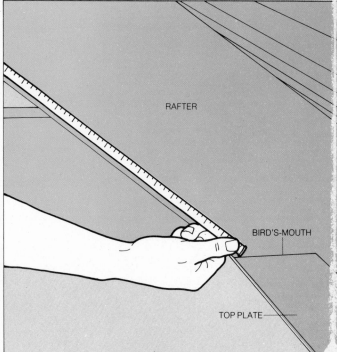

RAFTER

BIRD'S-MOUTH

TOP PLATE

2 **Cutting the ridge end.** Hold the T bevel near the end of a board of width and thickness similar to the old one but slightly longer. Place the T bevel handle against the edge of the board and the blade pointing away from the end. Draw a line along the blade and cut along the line.

3 **Cutting the eave end.** Measure along the rafter from the bottom of the ridge beam to the beginning of the bird's-mouth notch. Mark this length on the shorter edge of the new board. Measure with the T bevel and trim the new rafter to match the horizontal cut of the bird's-mouth.

73

4 **Installing the new rafter.** Hold the new rafter flat and poke the eave end between the sheathing and the top of the wall; lift the upper end of the rafter past the lower edge of the ridge beam. Twist the rafter into position and slide it next to the damaged rafter. If you have difficulty slid- ing the new rafter into place, trim it with a saw. Force the rafter up against the sheathing and against the ridge beam while a helper secures it to the old rafter with two or three nails. Drive four tenpenny nails through the ridge beam and into the end of the new rafter. Close up any gaps between the rafter and the sheathing with shims driven between the rafter and the top plate. For extra security if you are about to reroof, drive a nail up through the sheathing to locate the new rafter on the roof. From the outside, nail every 6 inches along the new rafter.

Sisters for a Sagging Rafter

Raising the sheathing. Hold a block of wood up against the sagging sheathing, next to the bowed rafter, and strike the block with a hammer to lift the sheathing slightly. Continue to force the sheathing up until it is straight. If necessary, insert shims between the sagged rafter and the sheathing to hold the sheathing in position temporarily. Nail 2-by-4 sisters to each side of the bowed rafter to bridge the sag and support the sheathing flat. Then put a locator nail through the sheathing, climb onto the roof and nail the sheathing to the 2-by-4s at 6-inch in- tervals. Caution: This procedure will cause leaks; use it only when you are about to reroof.

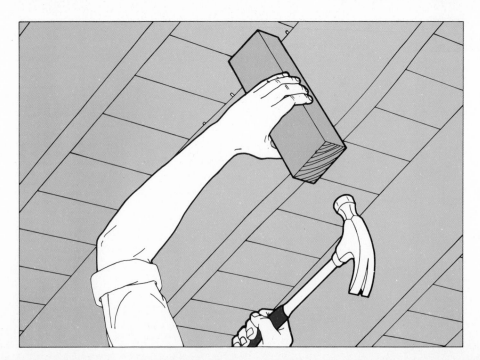

Replacing a Rafter

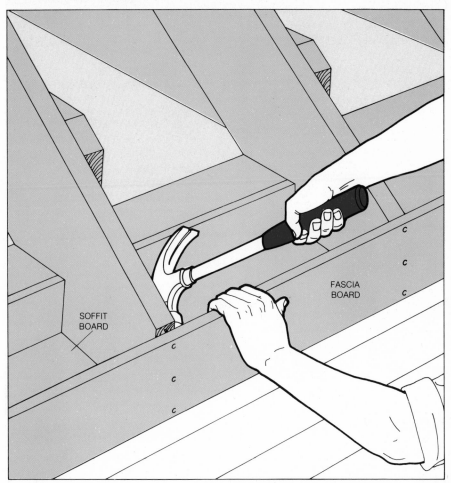

SOFFIT
BOARD

FASCIA
BOARD

1 Removing the rafter. Take off the sheathing. If
the soffit and fascia are held by common nails,
strike the inside faces of the fascia and soffit
boards with a hammer to pop the nailheads,
then remove the nails. If the boards are held by
finishing nails, drive them all the way through
with a thin nail set. Saw the damaged rafter in two
and work the pieces away from the ridge
beam and the top plate of the wall. Handle the
wood carefully to avoid splintering; this rafter
will be used as a template for making a new one.

2 Cutting the new rafter. Place a board the same
thickness and width as the old rafter but
slightly longer on the ground or attic floor. Piece
the old rafter together atop the new. Trace the
angles for the ends and cut the new rafter. If the
new wood is not quite as wide as the old
rafter, place the old rafter pieces so that their top
edges are flush with an edge of the new board.

Place the right angle bird's-mouth cut over the top
plate and lower the top of the rafter into position against the ridge beam. Toenail the rafter to
the top plate with four tenpenny nails—two staggered on each side—and end-nail it to the
ridge beam (*page opposite, Step 4*).

Asphalt Roofing—Handsome, Durable, Popular

The roofing material most widely used in the United States and Canada is asphalt shingles, which are generally made of wood- and rag-fiber felt impregnated with asphalt and coated with fine gravel. Most are strips 3 feet long and 1 foot wide. Two 5-inch deep slots—called cutouts—divide each strip into three 1-by-1-foot sections called tabs. Most come with dabs of adhesive across each strip just above the cutout tops. When the adhesive is softened by the heat of the sun, it seals the tabs of the overlapping shingles against strong winds and heat-caused curling. Two other types of asphalt roofing are roll roofing, which comes as rolled sheets of asphalt-impregnated felt coated with fine gravel (pages 82-83); and built-up roofing, which consists of alternating layers of hot tar and asphalt-impregnated felt (pages 84-85).

You can get shingles in a variety of weights. Generally, the heavier the shingle the longer it will last. An exception is the type made with a base of fiberglass felt instead of the usual wood-fiber base; it is relatively light but as durable as heavier ones made of wood-fiber felt.

Some textured shingles are constructed in two layers; the top layer has large cutouts to create the random effect of wood shingles or shakes. Other styles have lower tab edges cut at varying angles to provide an alternative to the more familiar pattern of horizontal, or course, and vertical, or bond, lines.

Roofs covered with standard shingles can be made more distinctive by adding ribbon courses—three shingles thick at the bottom edge—to emphasize the horizontal element every five to eight courses (page 81). One way to determine whether ribbon courses would improve the appearance of a roof is to photograph it, then emphasize every sixth course in the picture with a felt-tipped pen.

The colors of shingles can vary from one production run to another; be sure to buy all the shingles you will need at the same time and in packages marked with the same lot number. To keep shingles from becoming discolored in hot, humid climates, buy those that are specially treated to prevent alga and fungus growth. Store bundles where they will not get wet or be exposed to the sun.

If you have a worn-out asphalt roof with course lines 5 inches apart, reroofing generally is simply a matter of butting the tops of the new shingles up against the bottoms of the old (following pages). You must first tear off the hip and ridge shingles, nail down any shingles that have buckled and replace badly deteriorated or missing shingles (page 18) to provide a uniform surface. Install new flashing (page 17) and sweep the roof clear of loose granules and debris.

If your old roofing is not asphalt shingles, or has any kind of still-older roof underneath, tear off all the old roofing (page 68) and start with a bare deck. Remove only one section at a time. Install drip edges and new underlayment (page 70); if the underlayment gets rained on, give it at least a day to dry out. Mark new course lines on the roof with a chalk line (page 71). Reroof much as you would over existing asphalt shingles, but use the chalk lines rather than the old courses to align the new shingles. For the starter course, lay the shingles with their cutout edges facing the ridge.

Trim shingles wth metal shears or by scoring the backs with a utility knife and bending them. Fasten them down with large-headed galvanized or aluminum roofing nails (below) or with pneumatically driven staples (opposite).

Work across two or three full shingle widths from one corner, varying the width of the first shingle in each successive course to create a diagonal, stair-step pattern (page 78). Shingling in this way will minimize problems with color blending, since colors can vary slightly even between bundles with the same lot numbers. In areas where windstorms are common, factory-applied adhesive may not be enough to keep shingles flat on extremely steep slopes; add a 1-inch dab of roofing cement underneath each tab.

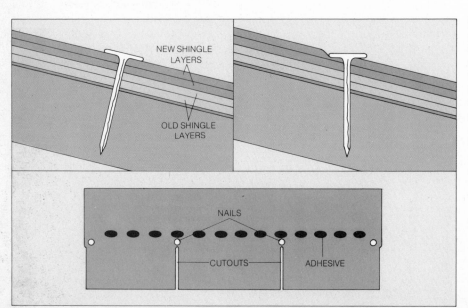

NEW SHINGLE LAYERS

OLD SHINGLE LAYERS

NAILS

CUTOUTS   ADHESIVE

Nailing shingles. Drive roofing nails just flush with shingle surfaces (far left). Be careful not to drive them crooked (near left), breaking the surface. Since the nail shank should penetrate through plywood sheathing or ¾ inch into board sheathing, reroofing over old asphalt shingles with standard material requires 1½-inch nails; for a new roof or one from which old shingles were torn off, use 1¼-inch nails. Heavy-textured shingles require nails 2 inches long. Fasten each shingle with four nails placed along a horizontal line ⅝ inch above the tops of the cutouts: one nail above each of the two cutouts and one nail 1 inch in from each end (left, bottom). Drive an extra nail beside any nail that does not sink into solid wood.

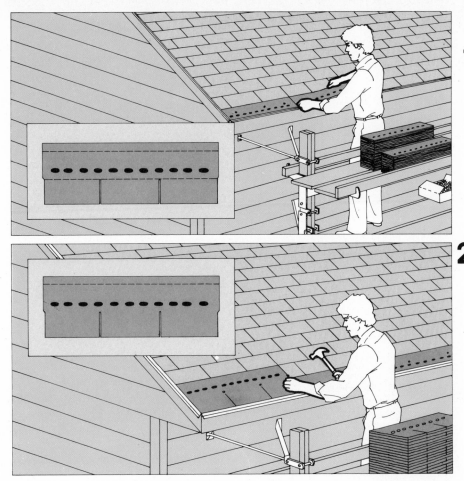

Installing Asphalt Shingles

1 Laying a starter course. Nail starter strips over the first course of old shingles, using three nails placed halfway up the strip, one in the middle and two 4 inches from each end. Make the strips 5 inches wide and 3 feet long by cutting shingles even with the tops of the cutouts, eliminating the flaps, and cutting again 5 inches above the cutouts, as indicated by the dashed lines in the inset drawing; in addition, trim 6 inches off the end of the first 5-inch strip so the joints between starter course shingles will not fall directly below the joints of the first course.

2 Laying the first course. Cut 2 inches off the top of two strips *(dashed line, inset)*. Nail them over the starter course and the second course of old shingles so that the top edge of the new shingles are butted up to the lower edge of the third old shingle course.

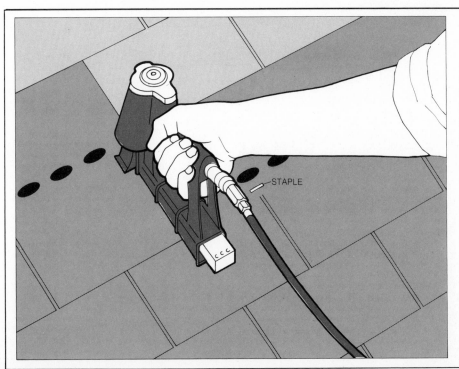

STAPLE

A Power Stapler for Speed

A pneumatic stapler powered by an air compressor—both available from rental agencies—can speed new roofing work by firing special staples to fasten shingles. Staples—with heads horizontal—should be placed in the same locations as nails *(opposite)*. Make sure that the staples and stapler are matched to your roofing work. Applying asphalt shingles to a bare roof requires 1¼-inch staples, which are often plastic-coated to give the staple legs extra gripping power. Reroofing over old shingles calls for 1½-inch staples; some of the smaller staplers will not hold staples this long. Even the larger tools may not be suitable for reroofing with heavy-textured shingles, which can require fasteners as long as 2 inches. When you rent the stapler be sure to get one with enough rubber air hose to allow you to reach all parts of the roof.

3 **Laying courses two to five.** Cut a half tab—6 inches—off the left end of a shingle. Align the cut end with the rake edge of the existing shingles and butt the top edge with the lower edge of the fourth old shingle course; the new shingle should then conceal all but 3 inches of the first new course. Nail the shingle down and add another shingle in this course, bringing it to within half a tab of the right edge of the first course. Install the third, fourth and fifth courses similarly: start each with a shingle cut a half tab shorter at the rake edge than the first shingle of the course just below, butt the shingles to the courses of the old roof, and extend each course to the right to within half a tab of the preceding course, creating a pattern of stairsteps going up to the left.

4th OLD COURSE

2nd NEW COURSE

4 **Laying courses beyond the fifth.** The first shingle of the sixth course (right) is just a half tab. The seventh course starts with a full shingle, untrimmed like the first, and subsequent courses through the 12th are started with trimmed ones, like courses two to six, to maintain the stair-step pattern. At this point the stairsteps will have reached the rake; return to the eaves and start a second stairstep with two full shingles, repeating the process until the roof is covered. Adjust spacing between shingles to keep cutouts aligned vertically. Trim the topmost shingles even with the ridge.

6th NEW COURSE

5 **Marking a valley.** Snap chalk lines onto new valley flashing, one on each side of the valley center line. Make the lines 6 inches apart at the ridge, but angling outward so that they diverge 1 inch every 8 feet toward the eaves.

Shingling a Hip

1 **A guideline for hip shingles.** Since the ridge shingles *(page 80)* overlap the hip shingles, install the hip shingles first, aligning them with a chalk line that is snapped alongside two tabs—thirds of a whole shingle—positioned at the peak and the eaves. The aligning tab at the bottom of the hip is permanently attached with two nails that are placed 2 inches above the adhesive strip and 1 inch in from the edges. The second tab, at the top of the hip, is not nailed but temporarily placed in position while you snap the chalk line next to one edge of the tabs. Shingle up the hip from the eave, using single tabs and aligning them with the chalk line and leaving 5 inches of each exposed. Trim the top tab even with the ridge.

6 **Shingling a valley.** Add shingles to the new courses on each side of the valley so that the last shingle of each course completely overlaps the chalk line on its side of the valley. If the top corner of the last shingle falls short of the line, nail one tab—a third of a shingle—next to the previous shingle in that course, and then install the last shingle as shown above. To avoid puncturing flashing metal, place the end nail of each course just short of the flashing edge. Shingle the valley to the ridge, snap a chalk line directly above the first and cut the shingles along the line with metal shears.

2 **Finishing hips.** Nail a tab that has been slit up the middle 4 inches to the ridge with the slit portion over the tops of the hip shingles. Overlap the slit edges to draw the tab tight to the hips *(above)*. Secure the overlap with a roofing nail.

Shingling a Ridge

Shingling a roof ridge. Ridges without hips are shingled in from each end toward the center. First nail a shingle tab at each end with the outer tab edge extending ½ inch beyond the rakes to align with the ½-inch overhang of the roof shingles. Snap a horizontal chalk line along the ridge, level with the lower edges of the two end tabs. Shingle in from the ends covering 7 inches of each tab with the succeeding tab.

On roofs with hips, cover the hips first as described on page 79, then cover the ridge as described above.

Finishing the ridge. Overlap the top portions of the tabs where they meet in the middle of the ridge; trim the top of the last tab installed on the left side so that it leaves at least 5 inches exposed on the last tab on the right side (*right*). Conceal the overlap with the lower half of a tab secured with a nail in each corner. Dab the exposed nailheads with roofing cement.

Shingling a dormer ridge. Begin at the outer end of the ridge and install ridge shingles as shown at the top of this page. When you reach the valley end of the ridge, trim the tabs to match the slope of the valley. Cover the end of the last shingle with the lower portion of a tab cut to fit the valley (*right*) and nail on both sides of the ridge. Dab the exposed nailheads with roofing cement.

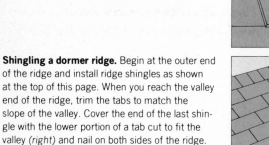

LOWER HALF OF TAB

CUT TAB

FULL TABS

Playing with the Patterns

Breaking on thirds. Install the starter strip and first course as on page 77. Start the second course with 4 inches—one third of a tab—cut off the first shingle *(dashed line).* Cut 8 inches—two thirds of a tab—off the end of the first shingle of the third course. For the fourth course, start with a full shingle.

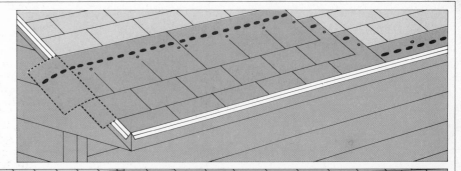

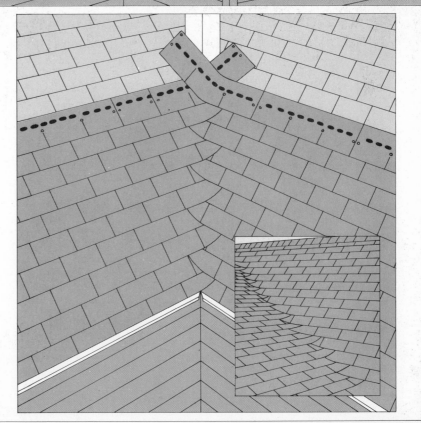

8″ STRIP   4″ STRIP

Forming ribbon courses. After six new courses are installed in the standard manner, cut and lay a strip 4 inches wide with its bottom edge ¼ inch below the top of the cutouts in the sixth course. Nail the remaining 8-inch strip over the 4-inch strip with its tabs facing uphill and its lower edge flush with the 4-inch strip. Cover both 4-inch and 8-inch strips with a course of full shingles, aligning their bottom edges with the edges of the cut strips. Repeat the procedure every six courses to create the shadow pattern illustrated above at right.

Weaving a valley. Lay shingles alternately from each side of the valley across the valley and at least 12 inches up the other slope of the roof. Space the valley-crossing shingle as on page 79, Step 6, to prevent joints between shingles from falling within 12 inches of the valley center line. Nail each shingle in both the standard locations *(page 76)* and in the upper corner of the end tab, but omit nails within 6 inches of the valley center line.

Where the slopes of the two valley sides differ *(inset),* cross with as many as three shingles from one side before crossing with a shingle from the other side to keep the seam between the two sides near the center of the valley.

Roll Roofing for Economy

Superficially, a scrap of roll roofing resembles asphalt shingle: like the shingle, the roll material consists of asphalt-impregnated felt—wood and rag fibers—covered with mineral granules. There the resemblance ends. The roofing comes in rolls that weigh between 50 and 90 pounds and unroll into strips 36 feet long and 3 feet wide. What is more, roll roofing has only half the life of shingles and lacks their esthetic appeal. But it is the fastest, easiest and least expensive roofing to install.

Roll roofing is especially suitable for low-slope roofs. It can be applied to roofs with pitches as little as an inch per foot—half the minimum for shingles—and because the surface of low-slope roofs is not normally visible, the plain appearance of the roll material usually does not matter.

The double coverage, concealed-nail application method shown on these pages gives the neatest appearance, the greatest protection and the longest life. To follow it, use roofing with a 19-inch selvage—the material that is hidden when the roofing is installed.

For the best job, remove the old roof covering, repair the deck if necessary *(page 68)* and install drip edges along the eaves and rakes *(page 70)*. However, you can reroof over roll or shingle roofing if you nail down loose or torn material and repair blisters. Unroll the roofing and cut it into pieces slightly longer than the roof. For a roof longer than 20 feet, it may be more convenient to cut shorter lengths and combine the pieces on the roof with lap joints. In either case, pile the strips near the top of the roof—on a day when the temperature is over 45°—and let them flatten out for a day or so. During the job, move each strip down for installation; when you near the top of the roof, move the pile down.

1 Laying the starter course. Cut a starter strip 19 inches wide from the selvage portion of a roll, and fasten it to the sheathing with roofing nails spaced a foot apart in three horizontal rows. Locate these nails 1½, 8 and 14½ inches up from the lower edge of the strip, and start them an inch in from the rake. The lower edge of the starter strip should overhang the eaves by ½ inch; trim the rake edges for a ½-inch overhang.

2 Laying the first course. Cover the starter course with a full-width, 36-inch sheet of roofing, which is trimmed after nailing to overhang rakes and eave ½ inch. Space nails a foot apart, in two horizontal rows, 4½ and 13 inches below the upper edge of the sheet.

3 Cementing laps. Fold the first course back to expose the starter strip and brush on asphalt cement (below). Unfold the first course and press it firmly into the cement with a stiff broom. Nail and cement additional courses until you reach the ridge.

4 Making vertical joints. To join sheets of roofing that are shorter than the roof, make vertical lap joints (below). Nail and cement a sheet in place, then drive a vertical row of nails an inch in from the end of the sheet with the nails spaced 4 inches apart. Install a second sheet, aligned to overlap the first by 6 inches, but do not nail through the overlap; instead, hold the second sheet back to expose the overlap, apply cement to the last 5½ inches of the first sheet, and press the end of the second sheet down into the cement. Drive a vertical row of nails into the selvage of the second sheet; locate the row 1 inch in from the lapped end and the nails 4 inches apart.

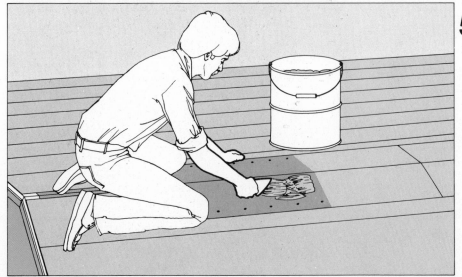

5 Finishing hips and ridges. Cut strips 12 inches wide and 36 inches long from a roll of roofing to make pieces to be nailed and cemented over hips and ridges (left). Use the 17-inch selvage of one of these pieces as a starter at the bottom of a hip or the end of a ridge. Snap a chalk line 5½ inches to one side of the hip or ridge; using the chalk line as a guide, fasten the starter in place with rows of nails 4 inches apart and an inch from the edges. Coat the starter with asphalt cement. Lap the granule-covered end of a second piece over the starter, nail the selvage as you did the starter and press the piece into the cement. Apply cement to the selvage of this piece and repeat the procedure for all subsequent pieces. Trim the selvage of the last complete piece to fit the end of the hip or ridge, then apply a granule-covered piece cut to fit the selvage. Lap the ridge over hips as shown on page 79.

Repairing a Built-up Roof

The most common covering for a flat or nearly flat roof consists of alternating layers of felt and tar, built up from a structural base. Installing such a covering is best left to professionals—the job calls for specialized equipment and the tar must be heated to temperatures as high as 500°. However, a homeowner can usually repair a built-up roof and coax several additional years of service from it before major repairs or replacement become necessary.

Blisters in the layers of the roof, collectively called the membrane, are easily repaired. They mark places where expanding water vapor has been trapped under or in the membrane, and may be the symptom of a leak in the roof; whatever their cause, they put a strain on the membrane that may lead to trouble. You can fix a blister that is small and isolated by the method shown below. Do not, however, try to repair large blisters or a roof that is covered with blisters.

Leaks in a built-up roof are a more serious matter. They usually occur at points where the membrane ends or is interrupted—at the edge of the roof, for example, or around chimneys and vent pipes—and can be stopped by the flashing repairs shown opposite. Leaks within the membrane are more difficult to pinpoint. On a roof that is not covered with gravel, try applying a coat of reflective asbestos-aluminum sealer, which can stop the leak—and, as a bonus, make the roof cooler in the summer. Otherwise, have the job analyzed by a professional.

Anatomy of a built-up roof. Many combinations of asphalt, tar, roofing felt, roofing paper, asphalt cement and roll roofing go into built-up roofs; the simple three-ply arrangement shown at right is perhaps the most common. A base layer of 43-pound "base sheet" is laid over the plywood sheathing of the roof and covered with hot liquid tar. Above the base sheet, courses of roofing felt alternate with layers of tar and the entire composition is covered with hot tar. On most roofs, a layer of gravel, often thick enough to add from 300 to 500 pounds to every 100 square feet of roof area, is embedded in the tar; on some the roof is finished with tar alone or with a layer of roll roofing.

Special flashing pieces, not shown here, are used at certain vulnerable spots. Vent pipes are completely enclosed by close-fitting flashing assemblies; where the roof meets a parapet wall, additional layers of roofing are brought to the top of the parapet and a cap flashing goes over both the roofing and the wall.

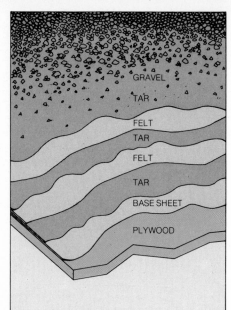

Patching a Blister

1 Removing the gravel. Using a stiff brush or a heavy-bladed roofer's tool called a spud bar, scrape the gravel from the surface of a gravel-covered roof about a foot in all directions beyond the blistered area. Caution: Gravel clings to a hot roof; in warm weather, do this part of the job early in the morning.

2 Installing a patch. Cut an X through the blister—but no deeper—with a utility knife, fold back the roofing and let it dry for an hour. Using a small trowel, coat the exposed area with roofing cement. Work the cement into the crevices at the edges. Smooth the flaps of roofing back into the cement. Cut two patches of roofing felt, one 6 inches and the other 12 inches wider and longer than the blistered area. Trowel cement around the area and embed the smaller patch in it; then cover the patch with cement and repeat the process with the second patch. Match the finish of the roof by pressing gravel or pieces of roll roofing into the topmost layer of cement.

Replacing Vent-Pipe Flashing

1 **Removing the old flashing.** Scrape away gravel 2 feet around the vent pipe, cut into the roof just outside the edges of the flashing base plate with a utility knife and slip a pry bar into the cut and under the edge of the base plate. Pry the base plate up, using a scrap of wood as a fulcrum to prevent damage to the roof membrane. Make patches of roofing felt to fit the gap you have made, with a round hole in each patch to fit over the vent pipe, and fill the gap with alternating layers of roofing cement and felt.

2 **Reflashing the vent pipe.** Trowel cement around the area of the vent pipe and, wearing gloves, fasten a new flashing assembly over the vent pipe with roofing nails driven through the corners and into the roof; crimp the soft metal of the flashing barrel into the vent pipe. Cut two sheets of roofing felt or patching fabric, one 6 inches and the other 12 inches larger than the base plate. Cut holes at the centers of the sheets to make them fit snugly around the barrel of the flashing assembly. Spread cement around the base of the assembly, fit the smaller sheet over the pipe and embed it into the cement; repeat the process with the larger sheet. Coat the top sheet with roofing cement and on a gravel-coated roof spread gravel over the cement.

Repairs at a Parapet

Installing a flashing patch. Scrape off any gravel within 6 inches of the damaged area—work carefully to avoid tearing the roofing material, which is especially vulnerable at a parapet— and, using a pry bar and a scrap of wood as a fulcrum, lift the nails, if any, that fasten the metal cap flashing to the side of the parapet. When the nails are completely free pry the entire section of the cap up and away from the parapet.

Coat the damaged area with asphalt roofing cement and embed in it a sheet of patching fabric 3 inches larger in all directions than the area to be repaired. Apply a second coat of cement and lay a second piece of patching cloth, 6 inches larger all around. Coat this patch and the area under the cap flashing with roofing cement, then bend the flashing down over the parapet and replace the nails and gravel.

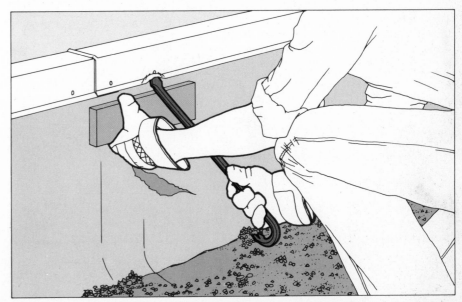

Classic Shingles and Shakes in Cedar or Redwood

Hand-split shakes and smooth-surfaced shingles serve as handsomely on the roof of a house as they do on the walls. Because of its cellular structure, the wood contributes a measure of insulation. And the overlapping layers of long, rigid cedar and redwood shakes or shingles reinforce the framing of the roof, making it more resistant to wind or hail damage.

For long life, both shakes and shingles need to drain quickly. Shakes require a minimum pitch of at least 4 inches to 12 inches horizontal run, shingles at least 3 inches to 12, but in some rainy climates professionals insist that the pitch should be 7 inches to 12 inches or steeper. Check a local supplier to find out what the practice is in your area.

Except in dry climates, shingles also need ventilation underneath to allow moisture to evaporate. If you are removing old roofing, you can replace solid sheathing with an open deck (page 69). If you are installing wood shingles over an existing layer of asphalt, first cover the roof with spaced ventilation strips (opposite, Step 3). Shakes are split so irregularly that they ventilate naturally, so you can put them directly over solid sheathing or asphalt shingles as well as on an open deck. But that same irregularity of surface will allow snow to filter underneath the shakes unless you interlay each course with felt strips (page 90).

Which material you choose and how widely you space the courses, or horizontal rows, depends on the roof.

To plan exposure accurately, divide the distance from the eaves to the ridge by the space you want to use for each course. Round off the result to the next highest whole number to determine how many courses you will need from eave to ridge. Then divide the number of courses into the original measurement to find exactly what exposure you will have to use.

Once you know the exposure and the dimensions of your roof (page 66), a roofing supply dealer can use charts to figure how many shingles or shakes your roof will require. Allow at least 10 per cent extra for trimming at valleys and projections such as chimneys, but be sure the dealer agrees to let you return unopened bundles. For shake roofs, also allow one roll of 30-pound felt 36 inches wide for each 120 square feet of roof surface.

Before you start roofing, build a portable roofer's seat (below) to make working more comfortable. The protruding nail points set in the base of the seat will keep it securely anchored wherever you want to set it on the roof, and will not damage the new shakes or shingles. A toeboard (page 89) makes it easier to walk across the roof and also provides a place to stack shingles or shakes.

Cedar and redwood are decay resistant, but a shaded roof in areas of prolonged heat and humidity may possibly need the added protection of a clear penetrating wood preservative.

Setting the Maximum Exposure of Each Course

Length of material	Roof pitch 3″ in 12″ to 4″ in 12″	Roof pitch 4″ in 12″ or steeper
Shingles		
16″	3¾″	5″
18″	4¼″	5½″
24″	5¾″	7½″
Shakes		
18″	Not recommended	7½″
24″	Not recommended	10″

Sizing courses. To determine how much of each shingle or shake you can leave exposed in each course, you need to know the length of the material and the pitch of the roof. Longer exposures than those shown above will make the wood dangerously vulnerable to drying out, curling and splitting. Shorter exposures will make the roof last longer, but will require more shakes or shingles.

How to Sit Down on the Job

Making a roofer's seat. From 1-by-12 lumber, cut a base 18 inches long and a seat 15 inches long. Lay the base on the roof at right angles to the ridge and place the seat on top, aligning the front ends. Pick up the back of the seat until it is level. Then cut a board to fit between the seat and base, and nail it in place at a right angle to the seat. Nail the front of the base and seat together. Cut four strips of ½-inch plywood 3 inches wide and 11½ inches long. Drive three 1-inch roofing nails through each strip. Turn the seat over and, with the roofing nail points up, space the strips evenly along the base. Nail the strips securely. When shingling, place the base on the roof, nail points down.

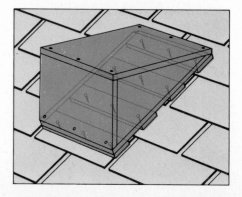

Installing Wood Shingles

1 **Trimming old roofing.** Remove the asphalt shingles along the ridges and hips, following the instructions on page 68. Trim off the asphalt shingles that overhang eaves and rakes with tin snips, or if you have one, a shingler's hatchet. To use it, grasp the handle of the hatchet with one hand and the knurled poll with the other hand. Then hold the sharpened heel of the hatchet snugly against the edge of the roof or the fascia and pull the heel along the eave to slice off the overhang.

2 **Preparing valleys, ridges and hips.** With eight-penny galvanized nails, fasten a pair of 1-by-3s spaced ½ inch apart over each side of the flashing in each valley. Then attach a pair of 1-by-3s along each side of all ridges and hips. Extend the ridge boards ½ inch beyond rakes, and make the ends of the boards flush with the ends of the ridges where they meet hips. Space each pair of boards ½ inch apart, but butt the top boards together where they meet over the ridge.

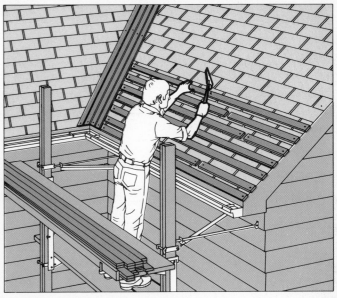

3 **Installing ventilation strips.** Using 1-by-3s if the exposure is 5½ inches or more, but 1-by-2s if it is less, nail the first ventilation strip parallel to—and flush with—the eaves. Extend the strip ½ inch beyond the rake edges. Measure up the roof from the lower edge of the strip and make a mark at a distance equal to the desired exposure. Nail the next strip parallel to the first, with its center line on the mark. Using the same spacing, nail parallel strips up to the ridge boards. Attach drip edge over the rake ends of the strips and install aluminum flashing in the valleys, using the methods shown on pages 16-18.

4 Laying the starter course. Position a shingle at each end of the roof with the butt overhanging the eave by 1½ inches and the rake by 1 inch. Using hot-dipped sixpenny galvanized nails, secure each shingle by driving a nail into each side 1½ inches farther from the butt than the distance for the desired exposure. If one end of the roof terminates in a valley as shown here, follow the techniques in Steps 5 and 6 to shape and install the shingle. If the roof terminates at a hip, install the shingle with the outside edge overhanging the hip by about ½ inch and trim the edge to the line of the hip.

Drive tacks into the butts of both end shingles and stretch string between them. Then, using the string as a straightedge, nail shingles spaced ¼ inch apart across the eave. Finish the starter course by covering the shingles with a second layer, aligning the butts and offsetting the vertical joints between shingles by at least 1½ inches.

5 Shaping valley shingles. Line up the butt of a wide shingle parallel to the eave with its bottom corner touching the center of the valley flashing. Place a narrow shingle with its inner edge in the center of the valley. Use the outer edge of the top shingle as a guide to draw a line on the bottom shingle. Saw off the inside corner of the bottom shingle and use the shaped piece for a pattern to mark and cut enough shingles to border the rest of the same side of the valley.

6 Installing valley shingles. Lay a 1-by-3 against the center of the valley. Set the shaped edge of the shingle against the board and drive a nail ¾ inch from the outside edge of the shingle, about 1½ inches above the desired exposure limit.

7 **Laying successive courses.** Lay overlapping courses of shingles up the roof, offsetting vertical joints by at least 1½ inches. Butt the tail ends of the last course against the top of the board at the ridge line. To gauge the exposure for courses up to 5½ inches deep, move the peg in the shingler's hatchet to the desired distance from the poll. Then hook the peg over the butt of the previous course and butt the new course against the poll. For exposures of more than 5½ inches, mark the hatchet handle to use as a gauge *(page 90, Step 2)*.

Shave shingles to fit around chimneys, vents and the like; then install new flashings as shown on pages 19-21.

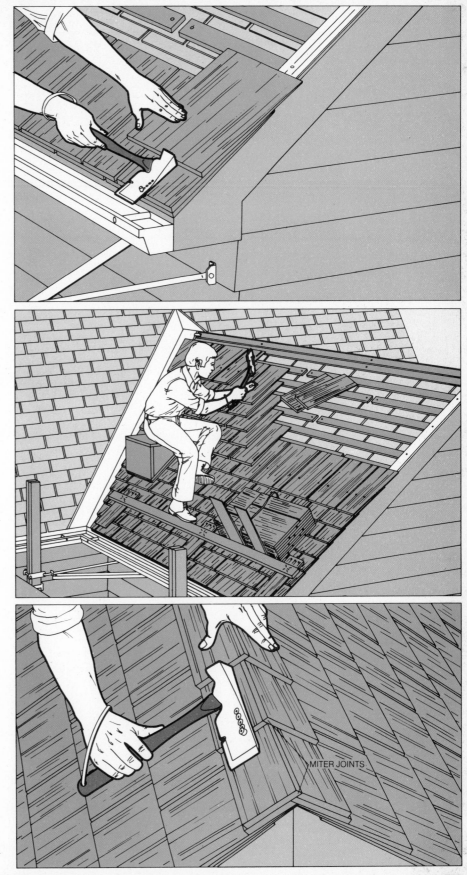

MITER JOINTS

8 **Making a toeboard.** After shingling as high as possible from a ladder or scaffold, use a toeboard made of a long 2-by-4 with shingles attached near the ends and at 3-foot intervals in between. The butt ends of the shingles, which should project upward 4 or 5 inches, are nailed into one of the newly installed courses. To support a bundle of shingles, wedge upright shingles into the gap between the top edge of the board and the course above. Install additional toeboards at 3- or 4-foot intervals as you work up the roof, leaving them to be pried off when the shingling is completed.

9 **Finishing ridges and hips.** Aligning the butt with the roof edge, set a factory-made ridge shingle at one end of a ridge and spread or compress the sides of the shingle to fit snugly. Place another ridge shingle with an opposite facing miter joint on top, then drive two tenpenny nails into each side of the shingles about ¾ inch above the bottom edges and at a distance from the butts equal to 1½ inches more than the exposure used for roofing courses. Using the hatchet peg or handle tape as a gauge and alternating miter joints, install an overlapping row of single ridge shingles across the roof. At the opposite end, cover the tail of the last shingle with a second unit, turned butt end outward.

For a hip, follow the same procedure but start from the bottom and work upward. Where the hip meets a ridge or another hip, shave the adjoining edges of the top hip shingles to fit snugly.

Installing Shakes

1 Laying the starter course. Roll out a 36-inch strip of 30-pound felt across the roof, align the bottom edge to the eave, and secure the top edge with roofing nails. Install a double starter course of shakes following the techniques used for shingles *(page 88, Step 4)*, but in this case let the butts extend 2 inches beyond the eave and the outside edges extend 1½ inches beyond the rake. Then cut a roll of felt in half lengthwise and nail an 18-inch strip above the starter course with its lower edge positioned at a distance above the butt equal to twice the desired exposure for the shakes.

2 Laying successive courses. Using a tape wound around the hatchet handle as a gauge for the desired exposure, install a course of shakes over the 18-inch felt strip. Then nail another felt strip over the tails of the shakes and repeat the procedure, interlaying each course with felt, up the roof. Install shakes at valleys and finish the ridges and hips as shown on page 89, Step 9.

Two Roofs of Yesteryear

With a few exceptions—one of them appears on page 64F—roofs of heavy earth or light plants turn up nowadays in historical books and movies. How were such roofs built and how did they work?

The plant roofs, at least, worked surprisingly well. If a raindrop falls on a reed held at an angle, the water will travel an inch or two along the reed before continuing its fall. If the reed is one of a bundle, the drop will hit another reed and run farther before it drops from the bundle. If the bundle is thick enough, no raindrop will get through it. If bundles overlap, as shingles do, the water will flow along the reeds clear to the eaves. In short, fibers that are round in cross section, apparently the worst of materials for holding out water, can do so if they are properly assembled and laid.

Such material, called thatch and consisting of grass, brushwood, bracken, heather or straw, bundled and piled on rude arbors, may have formed the earliest roofs built. Thatch still is the most practical roof in many parts of Africa, Asia, South America and Polynesia. It served the peasantry of Europe through the ages, and to this day, in the words of a professional English thatcher named Ernie Bowers, it provides "a marvelous roof, warm in winter, cool in summer."

Some years ago Bowers told of how he went about laying new thatching on a very old house in Norfolk. "I strip right down to the rafters before I start," he said. Bowers spent wintry days in the woods cutting straight hazelwood wands and splitting them (hazel, he says, is "the best splitting wood there is") to lay across the rafters as sheathing. For the thatch, he used 30-inch reeds cut in the Alde River where it meets tidewater—reeds that were "pickled a bit" in salt.

Bowers formed the reeds into bundles 14 inches thick, tied like sheaves of wheat. Working from a 48-rung ladder, he laid the first course, bundle beside bundle, with the tips of the reeds down. The butts of the bundles were clamped between the sheathing and a hazel wand, fastened to the rafters by nail-like reed hooks. The next course overlapped the first, hiding the wand. "Cruel work it is," says Bowers—the fibers of reed "play merry hell with your hands." Thatching

A thatching scene. In this copper etching from a 17th Century encyclopedia compiled for landed German gentry, one peasant scythes the straw for thatch, while others bundle, trim, stack, transport and lay it. The object of the demonstration, according to the authors, was to teach peasants to do the job without fatal accidents, for a dead peasant could thatch no more.

upward, he tilted his ladder parallel to the roof pitch and worked half prone, knees against the ladder's rails. He made the ridge of flexible straw, overlapped both ways. As a final touch, he cut the eaves off razor-sharp.

Some of the first settlers in the American West roofed their houses with a thatch of hay, but when they moved on onto the Great Plains they found little hay. In its place, they used a material that was literally underfoot and dirt cheap: sod, the source of the "Kansas bricks" of thousands of frontier houses.

One settler who built a sod roof was a 23-year-old printer named Howard Ruede, from Bethlehem, Pennsylvania, who laid claim to a homestead near Kill Creek, Kansas in 1877. Having learned how by watching his neighbors, Ruede set about building a half-dugout, a house

placed partly underground. He dug a chest-deep hole 10-by-14 feet into soil "sticky as putty." Choosing an area of buffalo grass, where wallowing buffalo had made shallow depressions that held water and grew tough-rooted grass, Ruede borrowed a sod-cutting plow to turn over long slices 12 inches wide and 2 to 4 inches thick. With a shovel he cut 18-inch lengths, and took them to the house site. "The sod is heavy and when you take 3 or 4 bricks on a litter or hand barrow and carry it 50 to 150 feet, I tell you it is no easy work," he wrote in his diary. He laid the bricks in a wall that was about a foot high around the hole.

Somewhere—his diary does not say—he acquired two 7-foot posts with natural forks at the top, and planted them in the centers of the short walls. For a dollar, a timber dealer some miles away cut and trimmed 20 feet of bur oak, 16 inches thick at the butt, and skidded it to Ruede's with a pair of Texas oxen. This might seem a heavy timber for a ridge pole; but every settler had heard of poles that broke under the weight of sod, at least once burying a man alive, and Ruede was playing safe. He and two friends lifted the pole into the post forks "by sheer strength of arm."

From a nearby lumber mill Ruede scrounged sawed rafters and boards. He sheathed the rafters with boards (most builders around Kill Creek had to settle for cornstalks), scattered straw on the boards, then, at last, he "covered the whole roof with a layer of sod, threw dirt on it"—to seal the joints—"and the 'House' was finished."

Howard Ruede's sod roof differed considerably from a thatched one in the matter of shedding water—to be blunt, it didn't. In theory, the sun would bake sod to waterproof adobe; in practice any heavy rain made sod leak; dirty water dripped inside long after the sun reappeared outside. Settlers fought this indoor rain, it might be said, with a raincoat over the bed, an oilcloth over the bread and an umbrella over the head.

Obviously, thatch, sound in principle and good for as much as sixty years, was the most serviceable roof of yesteryear. Sod, leaky and good for only five years, was no more than an unhappy expedient.

Slates and Tiles: Permanence at a Price

Slate and tile are among the most expensive of all roofing materials, but worth the cost if you want to put an old house back into mint condition or add elegance to a new one. Both materials require a roof with a slope of at least 4 inches in 12 and both can be applied over a single layer of asphalt shingles as well as over a freshly felted roof (page 70). However, both are heavy—slates average 7 pounds a square foot, tiles as much as 16 pounds; unless you are replacing old slate or tile and thus know the roof structure is made for the load, it may be wise to get a structural engineer to check your roof.

Standard roofing slate is 3/16 inch thick and 18 to 24 inches long. But it generally is sold in random widths that vary from about 8 to 14 inches, and most roofs are covered with assorted widths of uniform length. Slates come provided with nailing holes and irregularly sloped or beveled edges. When installing them, be sure the exposed bevel faces upward.

Roof tiles come in myriad sizes and shapes, but the most familiar is the single-barrel style shown on pages 95-97.

This tile is 13¼ inches long, 9¾ inches wide and ⅜ inch thick, and it is among the lightest in weight—at only 9 pounds a square foot.

Slates are laid with overlapping vertical layers. The maximum exposure or space between courses is determined by subtracting 4 inches (3 inches for roofs steeper than 6 in 12) from the total length of a slate and dividing the result in half. To plot regular courses, subtract 6 inches for crosswise slates at the ridge from the distance between the ridge and eaves. Divide this figure by the maximum exposure, round off the result to the next higher whole number, and then divide that figure into the distance again to find the exact exposure required.

Tiles, however, overlap horizontally as well as vertically so you must measure the width of your roof as well as its height to plot regular courses in both directions. For the barrel tile, the average length exposure is 10¼ inches and the average width exposure is 8¼ inches. Since you will need an assortment of specially shaped accessory tiles to finish the eaves, rakes, ridge and hips (pages 95-97), enlist the help of your roofing supplier to determine the exact exposures.

Installing slate begins with edging the roof with lath (Step 1), which cants the double-thick first course at the eaves and supports the outside edges of the final slates on the ridges and hips. Tiles require a 1-by-4 support along the ridges and hips (page 95, Step 1), and a 1-by-3 under the end units on the final rake (page 97, Step 7). Both slates and tiles are installed across the roof and up it, stair-step fashion, but while slates can be started from either the left- or right-hand end of the eaves, tiles must always be laid from right to left so they will lap horizontally.

Because of their brittleness, slates and tiles are nailed snug, but not tight. Professionals have special slater's gear (below) for cutting, shaping and punching holes in slates, and employ a water-cooled saw with a diamond blade to cut tiles. You can substitute a circular saw with a carbide blade for cutting, a bricklayer's hammer for shaping edges and a drill with a masonry bit for making holes.

Two Slater's Tools

A slater's hammer and roofing stake perform many jobs, and though they are expensive, they may be a good investment, particularly if the area to be slated is a large one. The forged one-piece hammer not only has a head for driving nails and a claw for pulling them, but also a point for punching holes and shear cutting edges along the shank for chipping slates to size. The crossbar of the stake functions as a straightedge.

But the chief purpose of the stake is to support a slate while it is being cut with the hammer as shown at right. After the point of the stake is driven firmly into a plank or scaffold board, the cutting line on the bevel side of the slate is positioned face down along the top edge of the stake. The crossbar then serves as a guide for the slater as he chips away the excess with measured strokes of the cutting edge of the hammer.

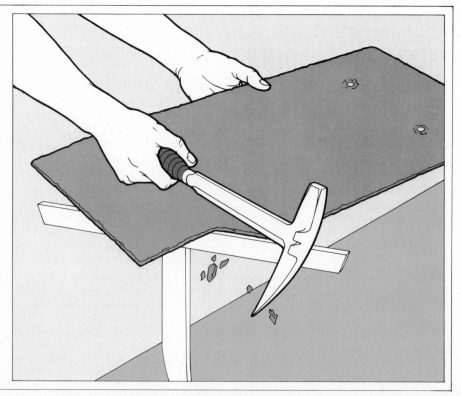

Installing a Slate Roof

1 **Preparing the roof.** After removing the ridge and hip shingles (*page 68*) and trimming the rake and eave edges, line the eaves and both sides of all ridges and hips with lath. Butt the edges on the strips running along ridges and hips.

2 **Laying the undercourse.** To ensure that the slates tip upward from the edge of the eave, install an undercourse of short slates. Pick out slates 3 inches wider than the desired exposure between courses. One at a time, lay each slate bevel side up and make an extra hole about 3 inches inside one corner on the solid end.

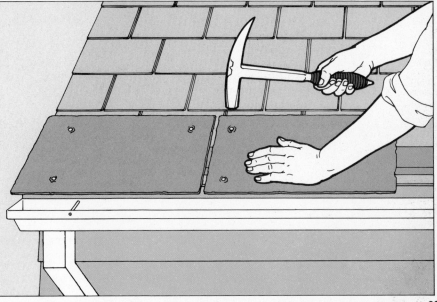

Install the first slate lengthwise—and bevel side up—at one end of the eave, letting one end extend ½ inch beyond the rake and the bottom edge project 1 inch beyond the eave. Secure the slate with threepenny slater's nails if it is up to 18 inches long, otherwise use fourpenny nails. Work across the eave, leaving vertical spaces about ¹⁄₁₆ inch wide between slates.

3 **Installing the first course.** Start at the end of the eave again and nail a full-length slate at right angles to the undercourse but aligned with its bottom and outside edges. Work across the eave, leaving 1/16-inch spaces between slates and selecting widths so that spaces between slates in the first course are offset by at least 2 inches from spaces in the undercourse.

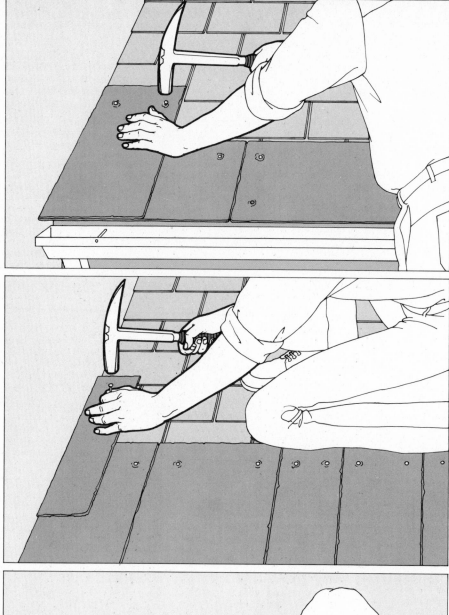

4 **Installing successive courses.** Begin the second course with a slate half as wide as the first slate in the first course. Position the slate so it extends 1/2 inch beyond the rake, and the butt is at the desired exposure distance from the butt of the slate in the first course. Work across the roof and up to the ridge, using full- and half-width slates alternately at the start of each course. Shape slates at valleys, using the techniques for cedar shingles (page 88, Steps 5 and 6). Be sure to offset all vertical spaces between slates in successive courses by at least 2 inches.

5 **Finishing ridges and hips.** For a ridge, butt the top edges of the final courses of slate on opposite sides of the roof. Then cut or select slates of a uniform narrow width, 6 to 8 inches, and lay them lengthwise along the ridge. Start at the lefthand end of the ridge and nail ridge slates to both sides of the roof, driving the nails through the vertical spaces between the final courses of slate. Overlap each successive slate, leaving about one third to two thirds of the preceding slate exposed. At the right-hand end of the ridge, punch or drill a hole into the outside end of the final ridge slates and make matching holes in the roof slates so you can secure the ridge slates firmly. Cover the exposed nails and the joint between the tops of the slates with roofing cement. For a hip, follow the same procedure, working from the bottom up. Where a hip meets a ridge, cut the final slates to fit snugly together.

Installing Single-Barrel Roof Tiles

1 Preparing the roof. Toenail a 1-by-4-inch strip set on edge to the tops of each hip and ridge, using sixpenny nails driven into opposite sides of the strip at 6-inch intervals. Where strips meet, as at the hips and a ridge, miter them to fit. Install metal drip edges along all the eaves *(page 70)*. Snap chalk lines to mark the horizontal courses and the first vertical course at the right-hand end, following the techniques shown on page 71. In this case, make the first horizontal chalk line 11¼ inches from the eave and successive lines at intervals equal to the desired exposure between courses. Make the first vertical line at a distance from the rake equal to the desired width exposure.

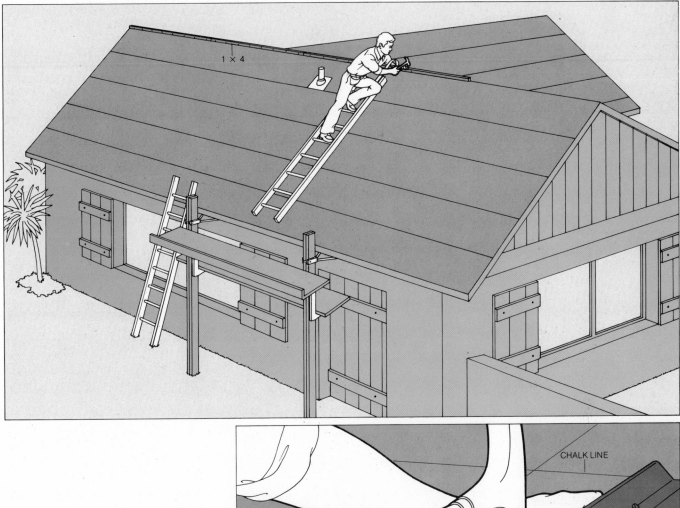

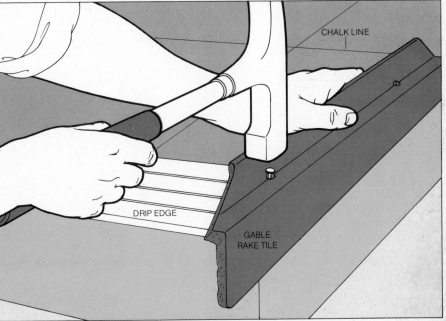

2 Laying gable rake tiles. Fit a gable rake tile snugly against the rake at the right-hand end of the eave. Align the top edge of the tile with the first horizontal chalk mark above the eave and let the bottom edge overhang the eave by 2 inches. Drive sixpenny copper or aluminum shingle nails into the holes on the corner of the tile. Install 3 feet of gable rake tiles, overlapping them to match the top edges with the chalk lines.

3 **Laying eave closure tiles.** Nail the first closure tile on top of the drip edge, flush with the eave and butting the first gable rake tile. To ensure proper spacing of the successive eave closure tiles, measure from the center of the installed tile a distance equal to the desired width exposure, and use a felt-tipped pen to make marks at this interval all along the drip edge. Adjust the width of the intervals to get a good fit at the left-hand rake. Install about 3 feet of eave closure tiles, centering each one on the drip-edge marking.

4 **Setting the first barrel tile.** Lay a bead of roofing mastic along the nail-hole line of the first gable rake tile. Center the barrel of the first "field" tile over the first eave closure tile, and press the curved right-hand edge of the field tile into the mastic on the gable rake tile to make a waterproof seal. Align the lip on the left-hand edge of the tile with the vertical chalk line. Then align the top edge with the first horizontal chalk line so the field tile projects 2 inches beyond the eave. Drive a sixpenny nail into the prepunched hole at the top of the field tile.

5 **Laying the first course.** Center the barrel of the second field tile on top of the second eave closure tile with the curved edge over the lip of the first field tile. Align the top edge with the first horizontal line. Nail the tile in place. Use this method to lay five or six tiles across the eave.

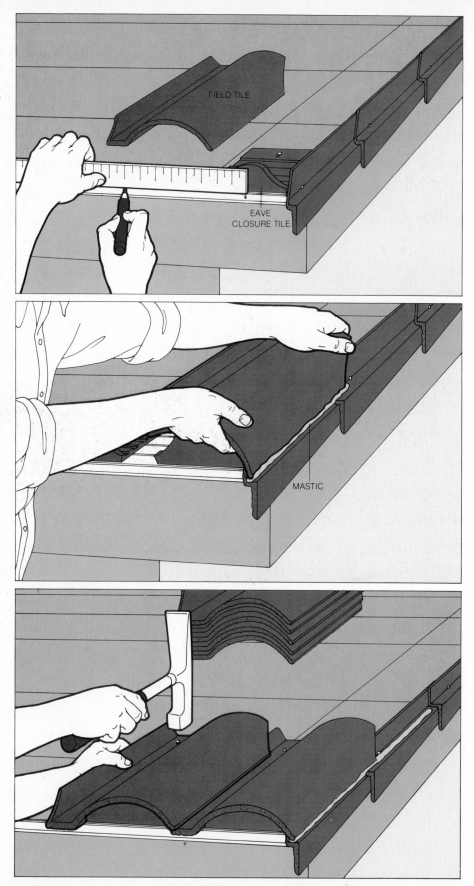

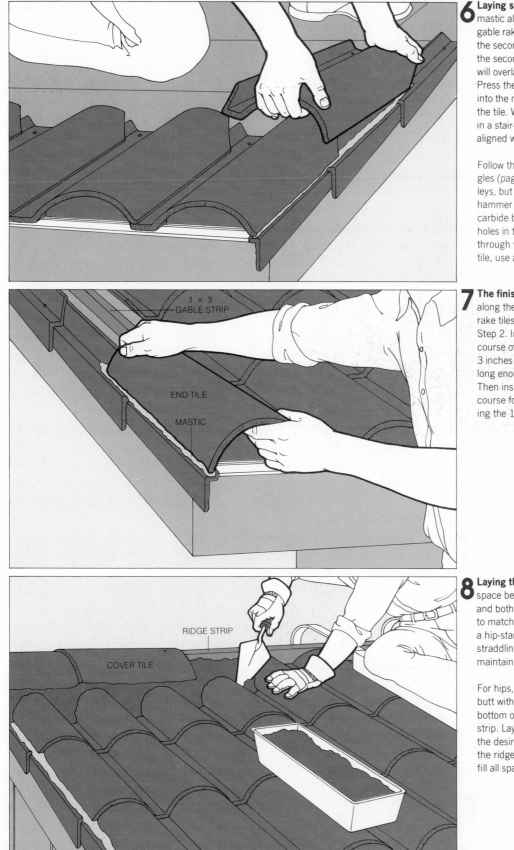

6 **Laying successive courses.** Lay a bead of roofing mastic along the nail-hole line of the second gable rake tile. Then position the first field tile of the second course with its top edge aligned to the second horizontal chalk line. Its bottom edge will overlap the tile below it by about 3 inches. Press the curved right-hand edge of the field tile into the mastic and drive a nail into the top of the tile. Work across the roof and up to the ridge in a stair-step manner, checking that tiles are aligned with the horizontal chalk lines.

Follow the techniques shown for cedar shingles (page 88, Steps 5 and 6) to install tiles at valleys, but shape the tile to fit with a bricklayer's hammer or with a circular saw equipped with a carbide blade. Use a carbide bit to drill extra holes in the tiles, if necessary. To avoid nailing through valley flashing when laying a cut field tile, use a cleat, as shown on page 17.

7 **The finishing course.** Before installing end tiles along the left-hand rake, lay left-hand gable rake tiles, following the procedure on page 95, Step 2. Install field tiles to within one vertical course of the rake. Toenail a 1-by-3, laid on edge, 3 inches inside the rake. The 1-by-3 should be long enough to extend from the eave to the ridge. Then install an end tile at the end of each course following the techniques in Step 4 and using the 1-by-3 as a nailing strip.

1 × 3 GABLE STRIP

END TILE

MASTIC

8 **Laying the ridge and hip tiles.** For ridges, fill the space between the top courses of field tiles and both sides of the ridge strip with mortar tinted to match the tiles. Beginning and ending with a hip-starter tile, lay hip-and-ridge cover tiles straddling the ridge strip. Overlap tiles to maintain the desired exposure.

For hips, cut the last field tile in each course to butt with the hip strip. Then, starting at the bottom of the eave, nail a hip-starter tile to the hip strip. Lay ridge cover tiles up the hip, maintaining the desired exposure. Where the hips meet the ridge, cut or shape the cover tiles to fit, and fill all spaces with mortar.

RIDGE STRIP

COVER TILE

New Modes in Metal Roofing for the Home

A sheet-metal worker is likely to tell you that the cat on a hot tin roof was, in fact, on a hot terne roof. Or possibly copper, galvanized steel or even stainless steel. And these days it probably would be aluminum. But never tin.

The metals actually used for roofing, all old standbys in some regions, are increasingly popular on new housing, especially of contemporary design. All are striking in appearance and, when properly maintained, durable. Some—particularly the ridge-seamed roofs made of terne or copper—require special tools and professional skills to install. But the old-fashioned "tin" roof of the American countryside—actually made of overlapping panels of corrugated galvanized steel—is easy to put up, and so is its modern offspring, the ribbed-panel roof of steel or aluminum (pages 100-101). Almost as simple to handle are the newest types, aluminum shakes and shingles.

All metal roofs—flat, ridged, ribbed or corrugated—are simple to repair. Leaks are stopped by coating with roofing cement or asphalt paint. On a flat or seamed roof (below), a hole or tear in terne, steel or copper can be patched by soldering on a new piece of matching metal; for aluminum (which cannot be soldered), or for any roof that has been coated with tar or roofing cement, a fiberglass patch is laid in fresh roofing cement. The repair of a hole in a panel, shingle or shake roof is even easier and more effective: simply remove the damaged unit and nail in a new one.

The differences among the metals used has a profound effect not only on cost and ease of installation and repair, but on appearance and durability. Each metal has its own properties and problems.

Terne, steel coated with an alloy of lead and tin, is easily painted. Indeed, its durability normally depends on paint: new terne metal should be coated on both sides with red iron-oxide linseed-oil paint, and exposed surfaces should get a second coat of oil-based paint (the second coat can be almost any color, a fact that makes terne a favored metal among modern architects). A more expensive version, terne-coated stainless steel, need not be painted at all. For repair, both varieties should be soldered with noncorrosive rosin flux.

Copper requires little maintenance and can be soldered with acid flux. You can, if you wish, prevent it from weathering to a greenish copper oxide by wiping the fresh reddish-brown surface occasionally with linseed oil. However, runoff from copper can stain light-colored surfaces below it, requiring frequent repainting.

Galvanized steel—that is steel coated with zinc—can be soldered easily and will resist rust until the zinc coating wears off. You can increase its durability by painting it with a zinc-rich paint after a year or two of exposure. Steel coated with aluminum or with aluminum-zinc combinations will resist corrosion even when unpainted, and stainless steel, the most expensive version, is all but maintenance free. Most steels should be soldered with a resin flux.

Aluminum's light weight and malleability make it a natural for roofing. It cannot be soldered, a problem solved by some manufacturers with a variety of clip-fastened or interlocking systems. Painting is unnecessary, but colored and embossed surfaces are available.

Flashing for all these metals should, if possible, be done in the same metal. If unlike metals must come in contact, as when galvanized steel is laid over existing copper flashing, the surfaces that touch should be coated with a bituminous-based paint; if the covering metal is aluminum, the paint should also contain aluminum pigment. Where an entire metal roof is covered with a different metal, separate the surfaces with layers of felt and rosined building paper.

A Soldered Patch for a Flat or Seamed Roof

1 **Shaping the patch.** Cut a patch at least 2 inches larger than the tear or hole, snip off the corners and turn the edges under ½ inch. Sandpaper the turned edge to a shine and coat it and the corresponding roof surface with flux.

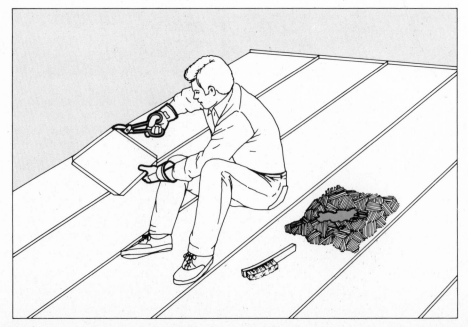

2 **Soldering the patch in place.** Set the patch in place and weigh it down with stones or bricks. Using the heaviest electric soldering iron available or a 3-pound torch-heated iron, heat the edge of the patch until solid-core solder flows freely into the joint, then work your way around all four sides. If you have used an acid-base flux, wipe off all residue with a wet rag.

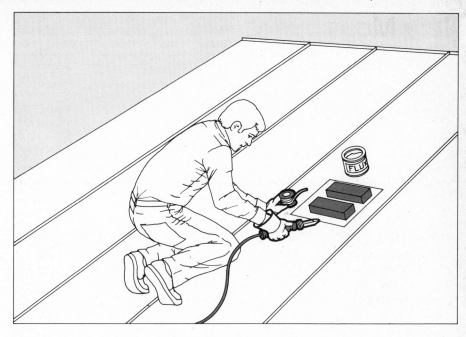

A Cold Patch

A cement and plastic sandwich. On a roof covered by tar or roofing cement, scrape or wire-brush the patching area to form a firm, flat surface, coat it with roofing cement and apply a fiberglass patch cut from a sheet or roll. Cover the patch with roofing cement, add a second patch and then cover the entire assembly with a final coat of cement.

Covering a Roof with Panels

The corrugated panel of galvanized steel, a century-old style of metal roof known for its ease and speed of installation, has spawned modern variations made not only of galvanized steel but also of aluminum or aluminum-coated steel and shaped in ribs as well as corrugations. Panels come in lengths up to 24 feet (or even longer, on special order) so that end lap seams are not necessary, and the newer types have overlapping side seams designed to prevent water from being siphoned up and under a panel.

All these roofs, old style or new, are installed by much the same method; the instructions on the following pages can be used for either type.

Most manufacturers sell accessories made to fit their products—nails prefitted with washers for leakproof fastenings, gable trim to accept the panels at the rake, ridge pieces and filler strips to seal edges at the eaves and ridge.

To get the tightest, most quiet roof, install the panels on a solid deck covered with roofing felt (pages 68-71). Normally, panels are precut to length by the supplier. If you must shorten one, cut through the antisiphon cap with a hacksaw, then complete the cut with tin snips. An entire stack of panels can be cut to length with a power saw. For extra protection against water seepage, caulk the panel seams, preferably with rope caulking laid along the edge of the antisiphon seams.

1 **Gable trim and filler strip.** Nail the manufacturer's recommended gable trim along the rake with washered roofing nails at 12-inch intervals, then tack a molded-rubber filler strip for the first panel near the edge of the eave. Do not nail the strip; it will be secured by the panel nails.

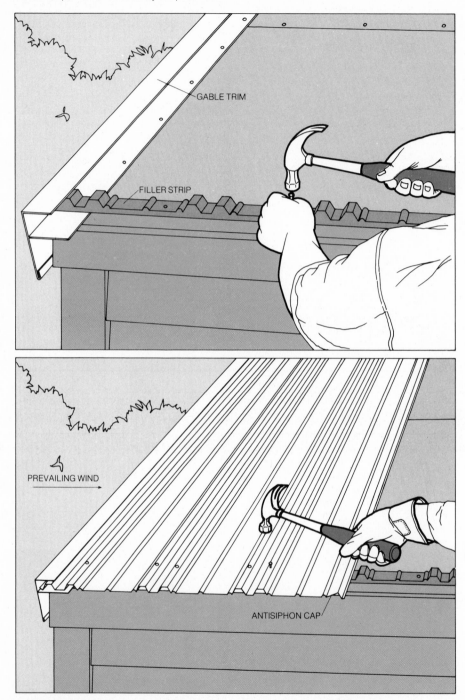

2 **Nailing the first panel.** Starting at the roof edge farthest from the source of the prevailing wind, lay the first panel flush to the ridge and overhang the eave by 2 inches, with the antisiphon cap facing into the prevailing wind. Nail through the ridges of the ribbing in the pattern recommended by the manufacturer, driving each nail just deep enough to compress its washer without dimpling the rib. Do not, however, fasten the final rib over the antisiphon cap—this rib will be nailed down when you install the next panel. Use caulking or the manufacturer's recommended sealer to plug any nail hole that is inadvertently enlarged.

3 **Lapping the second panel.** Lay the second panel with the first of its ribs overlapping the unfastened antisiphon cap. Nail this lapped seam at intervals of no more than 24 inches—on a solid deck drive the nails 12 inches apart. Complete the manufacturer's nailing pattern on all ribs except for the one that will be lapped by the next panel, and continue across the roof.

The one-rib overlap is suitable for both the double-rib panel shown here and the single-rib models offered by some manufacturers (inset). For corrugated panels, which do not have antisiphon caps, use a two-rib overlap.

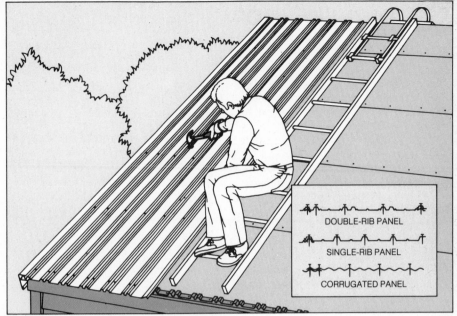

DOUBLE-RIB PANEL

SINGLE-RIB PANEL

CORRUGATED PANEL

4 **Cutting the last panel to fit.** Trim the last panel to fit inside the gable trim. Score the panel at the measured line with the corner of a screwdriver blade and snap it over the edge of a workbench or a firm, raised surface.

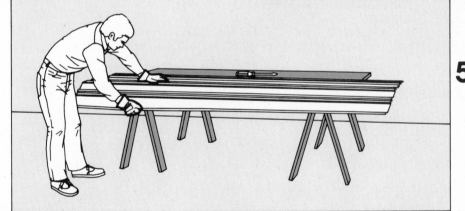

5 **Capping the ridge.** After installing the last panel, lay over the ridge a two-piece cap, which comes in sections molded to fit the pattern of the panel ribs and adjustable to any pitch. Cut away the lip of the gable trim to make room for the ridge cap at both gable ends; between the ends, lap sections of the cap at least 6 inches and apply a bead of caulking to the lap. Nail the cap down at every panel rib, and plug it at the ends with flexible closer strips provided by the manufacturer (inset).

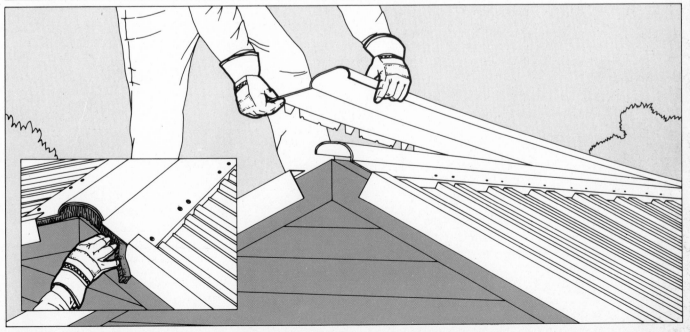

Raising the Roof

Working with angles. A graph-paper drawing of a dormer, drawn to scale, makes an indispensable guide to the angled intersections of the vertical, horizontal and sloping boards that will frame the structure. An L-shaped framing square marks the angles for the cut boards—from top to bottom, a side cut, a bird's-mouth, a bevel cut and a flush rafter *(pages 104-105)*—and an adjustable tool called a T bevel can be used to transfer all of these angles to other boards.

With a few changes, a roof can do more than look handsome and keep out the rain. Some roofs perform these functions well enough but block out desired light or form cramped, useless attic spaces; such roofs cry out for skylights or dormers. Other roofs—typically flat ones over garages—seem to ask for a secondary function as outdoor decks. And some areas next to a house may benefit from a roof where none exists, perhaps to make a carport or a porch. The following pages tell how you can use roofs better, either by putting up a new roof or by raising a roof line with an added structure.

A simple shed roof over a patio or a driveway *(pages 106-109)* adds usable space to your house and also introduces you to the craft of the house builder. The job calls for one of the builder's most useful instruments, the framing square *(pages 104-105)*. From its appearance the square seems suited only for measuring lines and marking right angles. In fact, it is designed to supply all the angle marks that a roof builder needs to make his rafters fit at ridges and walls as perfectly as the parts of a Chinese puzzle.

Raising a roof for a skylight *(pages 116-121)* lets sunshine flood a stairwell, a corridor or a windowless room. Similarly, a dormer *(pages 110-115)* can light an attic or a new room built in the attic. These are not new concepts: roof openings are as old as the builder's art, and dormers may go back a thousand years or more. According to legend, the dormer began in medieval Germany as a sly way to add bedrooms without adding taxes: German tax collectors assessed houses according to the number of their stories—up to the eaves.

An existing flat roof can be decked over *(pages 122-125)* to create new outdoor leisure space. Angles are the key to the deck's structure—you must build it over wedge-shaped joists that compensate for the roof's slight pitch—but once a pattern joist has been made the rest can be marked and cut rapidly.

To make a dormer, skylight or deck, you will have to cut through a wall or roof. Prepare for sudden rain with some heavy-duty polyethylene sheeting to tack over the opening. For the shed roof, skylight and dormer, use metal framing connectors—plates bent into angles and prepunched for the nails that tie the structural lumber together. Metal post anchors, caps and hangers were developed in the 1940s to provide strong fastenings in hurricane-prone areas—they were once called hurricane clips. They caught on elsewhere not only for their strength (half again as great as plain nailing), but for their ease of installation: they eliminate awkward butt-nailing and toenailing, and hold boards securely while you nail them in place. The skillful builder of an earlier day might have rejected these pieces of hardware, but they will make your job easier and better.

The Art of Cutting a Rafter to a Custom Fit

Because nearly all roofs are pitched—even most so-called flat roofs have a slight pitch—marking, cutting and installing rafters takes the craftsman out of the world of level, plumb and square. A rafter—unlike a joist or stud, which can simply be crosscut to length—may require tricky cuts at each end so that it can be mounted at the proper angle. Its higher end must be beveled to meet a ridge beam in a gabled roof or a wall-mounted header in a shed roof. The lower end needs either a horizontal bevel—for roofs with no overhang—or more commonly a bird's-mouth cut, a notch resembling the open mouth of a bird, to fit over and around the horizontal support.

To position and mark these cuts quickly and accurately, carpenters have long relied on a deceptively simple tool called a steel square or framing square. Essentially two rulers joined at a right angle, the tool makes it possible to cut a rafter when you know the run, rise and slope of the roof, three figures that must be deter-mined before any roof is built. The run is the horizontal distance from the outside of the wall, or building line, to a point directly below the ridge (below). The rise is the vertical measurement between the center of the ridge beam or supporting header and the level of the top of the walls. The slope is the rise of the roof, expressed in inches, for every 12 inches of run (formula, page 106). For example, a roof with a rise of 15 inches and a run of 5 feet has a slope of 3 in 12.

Use a ruler and a framing square first to get a quick estimate of the length of the rafter so you can buy the necessary lumber. Lay a ruler diagonally across the square so that it intersects the rise (inches read as feet) on the outside edge of one leg of the square, and the run (also inches read as feet) on the outside of the other leg. The distance between the two is the length of the rafter from the ridge to the wall. This figure determines the stock from which the rafters must be cut (chart, right). When you buy the boards, be sure to add the length of any planned overhang.

To mark the rafter cuts precisely, work down the rafter from the ridge end to the wall end. First mark the angle for the ridge cut (Step 1). Measure off the exact length of the rafter by stepping off triangles down the board (Step 2). Mark the bird's-mouth (Step 3) and finally the wall end, allowing enough extra beyond the bird's-mouth for any overhang (Step 4). After cutting one rafter, use it as a template to cut the remaining ones.

Choosing the Rafter Stock

Maximum Span	Board Size
5'4''	2 × 4
8'4''	2 × 6
11'0''	2 × 8
13'11''	2 × 10

Anatomy of a rafter. A rafter extends on an angle from ridge to building line. The ridge cut and the bird's-mouth notch allow the sloped rafter to fit against the face of the ridge beam, or header, and the top of the wall. The rafter's lower end is cut parallel to the house walls. the ratio between the rise and the run is the slope, which is generally expressed in inches per foot.

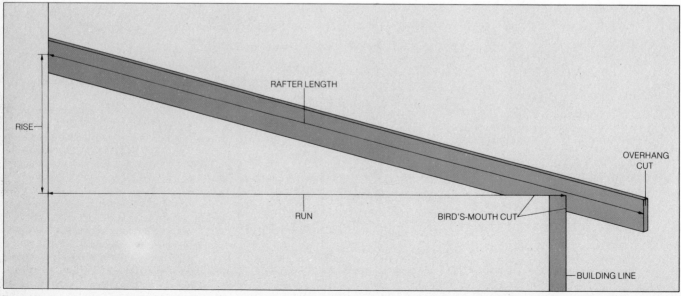

RAFTER LENGTH

RISE

RUN

OVERHANG CUT

BIRD'S-MOUTH CUT

BUILDING LINE

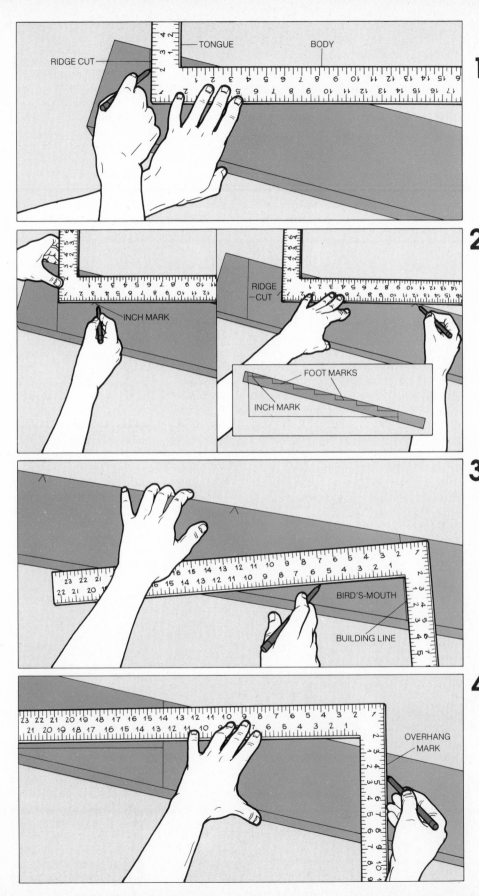

How to Use a Framing Square

1 Marking the ridge cut. Place a steel framing square near the ridge end of a rafter with the square's tongue close to the end of the board. Position the tongue to align the slope in inches (in this example, 3 inches in 12) with the top edge of the rafter. Position the body to align the 12-inch mark with the same edge. Draw a mark along the outer edges of the square then remove the square and extend the line across the width of the rafter.

2 Stepping off the length. To mark a rafter with a framing square, measure the run and mark off the inches first. Setting the square in the position described in Step 1, locate the inches part of the run on the body of the square (far left) and make a pencil mark on the rafter. In this example, the rafter run is 8 feet 3 inches, and the mark is made at the 3-inch point of the body. To mark off the first full foot, align the tongue of the square with the pencil mark, position the tongue and body as in Step 1, and mark the edge of the rafter at the 12-inch mark on the body (left). Repeat for each additional foot of run. Finally, set the square as if to mark off another foot, but use the tongue to mark the building line (inset).

3 Marking the bird's-mouth cut. To plan the complex rafter cut at the building line, where the rafter fits over the top of the wall, reverse the framing square and align the inside edge of the tongue with the building line. Position the square so that the inside edge of the tongue intersects the bottom of the rafter at half the rafter width (on a 2-by-6 rafter, which is actually 5½ inches wide, the tongue should intersect the rafter edge at the 2¾-inch mark). Mark both of the inside edges of the square with a pencil; the angular cut formed by the pencil marks is the bird's-mouth.

4 Establishing the overhang. After stepping off the length of the overhang by the method shown in Step 2, reverse the square as shown (left). Align the rise in inches, as it appears on the outside edge of the tongue with the overhang mark and align the 12-inch mark on the body with the top of the rafter. Draw a line down the tongue to mark the cut. Now cut the rafter along the line. Cut along the marks for the ridge, the overhang and the bird's-mouth. Because rafter lengths are based on the total run—no allowance is made for the thickness of the ridge beam—you will have an extra ¾ to 1½ inches for adjustment of cutting errors. After trial-fitting the rafter, cut off any excess parallel to the ridge cut.

Extending a House with a New Shed Roof

A single-pitched roof, the kind familiar on sheds and lean-tos, makes a simple but serviceable cover for a common home-improvement job: the roofing-over of an existing concrete slab, such as a patio or stoop, to convert it into an entryway, a carport or even an enclosed extra room. The shed roof is, in effect, half of a gabled roof, supported at its peak by the side of the house rather than a ridge beam, and at its eaves by roof posts resting on the concrete slab.

Most properly built slabs are strong enough to carry the weight of the roof, so that no foundation need be constructed. For an entryway or room addition, a satisfactory base is provided by a standard patio slab, normally 4 inches thick and reinforced with 6×6-10/10 wire mesh (that is, 10 gauge wire spaced 6 inches apart). Check a slab for thickness by digging a trench along the sides; you can assume that the reinforcement is adequate if the slab has withstood several winters without cracking or heaving. If you are converting a patio to a carport, which needs heavier reinforcement to support the car, consult the type of specialist usually listed in classified telephone directories under "Drilling and Boring Contractors"; for a modest fee, he will take a test core of the concrete to determine whether it will stand up to the additional weight of your car.

The directions on the following pages describe how to build a shed-roofed structure up to 14 feet deep—the only limitation on its length are the size of the slab and the length or width of your house. Start by drawing plans to scale, based on the drawing below and showing front and profile views of the slab and the house wall. These plans will help you to estimate materials and to determine the exact dimensions and shape of the roof; in most areas, they must form part of your application for a building permit. When drawing the existing wall, include obstructions such as windows, eaves or gutters, which can affect the placement of joists and rafters. Then, still working to scale, draw in the ceiling height—normally 8 feet 1 inch above the floor for additions and carports, 7 feet 1 inch for entryways. Next choose a height for rafters at the house wall. This height sets the slope of the roof, which for appearance should blend with the existing house lines but, in any case, must lie well below obstructions on the wall. If the rafter height you choose severely limits the slope of the roof, remember that the slope can in turn limit your choice of roofing materials (page 67).

Check the charts at right to find the proper lumber size to use for joists and headers, and draw these elements into your plans. If the headers are larger than the joists, reduce the span by adding one or more additional supporting posts. Cutting the space between posts reduces the distance the headers must span without support; for these shorter spans, use smaller header lumber the same size as the joists. Find the rafter length by the method used on page 104. Consult the same page to find the rafter size, and indicate these dimensions on your plans. All lumber for rafters, joists and headers should be Douglas fir or hemlock; softer woods, which are not as strong, must be larger for any given span and are not generally used for framing roofs.

Finally, determine the slope of the roof—the vertical rise in inches for each 12 inches of run—an essential element in calculating rafter cuts with a framing square (pages 104-105). This calculation, too, can be made from the plans: multiply the total rise from the top of the joists to the center of the rafter plate by 12 and divide this figure by the run or depth of the structure in inches. For example, a roof with a total rise of 30 inches and a depth of 10 feet (120 inches) would have a slope of 3 inches in 12.

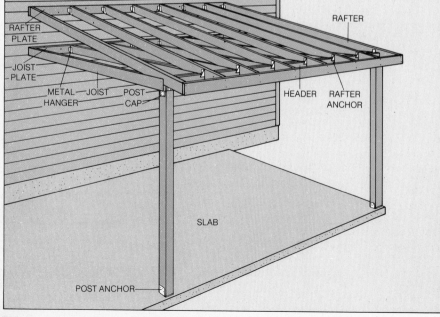

Anatomy of a shed roof. Horizontal joists and sloping rafters, both spaced at 16-inch intervals, define the ceiling and roof of the structure. At the outside wall, the joists and rafters are supported by metal hangers and anchors screwed to headers, which rest on corner posts; the posts are secured to the concrete slab with metal anchors and to headers with metal caps. At the house wall, the joists and rafters are fastened to plates also with steel hangers. The plates are secured to wall studs with lag screws.

For structures larger than 8 feet wide or 10 feet long, you should install two studs on the outside wall and add 2-foot-wide plywood panels for horizontal stability.

Joist and Header Stock

Joists

Maximum span	Board size
6'6"	2 × 4
10'	2 × 6
13'3"	2 × 8
16'9"	2 × 10

Headers

Maximum span	Board size
4'	2 × 4
6'	2 × 6
10'	2 × 8
12'	2 × 10

Building the Roof

1 Marking and measuring. Snap chalk lines on the slab to outline the structure. Where the lines meet the house, draw plumb lines up the wall to the approximate rafter height. Find the highest point on the slab at the wall and measure up the wall to the tops of the joist and rafter plates. At these heights, snap horizontal lines across the walls between the plumb lines. Cut a joist plate to match the distance between the plumb lines, and cut a rafter plate 3 inches longer.

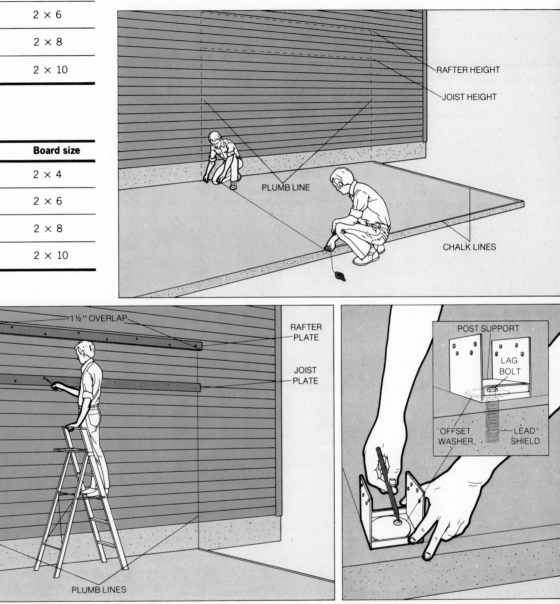

2 Installing joist and rafter plates. Fasten the plates to each wall stud with ⅜-inch lag bolts 5 inches long. The edges of the joist plate should be flush with the plumb lines on the wall. The edges of the rafter plate should extend 1½ inches beyond the lines so that the end rafters will, like the inner rafters, butt against the end plate. For brick or brick-veneer walls, drill ⅜-inch holes in the plates every 16 inches, position the plates and mark the holes' location on the wall. Drill ⅝-inch holes 4 inches deep at every mark and insert shields. Reposition the plates and screw them on with lag bolts. On a cinder-block wall, hang the plates from wing bolts inserted into the hollow sections of the block at 16-inch intervals.

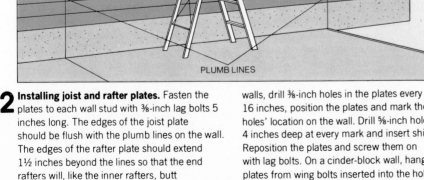

3 Installing post anchors. Align each anchor with an outside corner of the chalk lines, set the metal washer inside the anchor base and mark the outline of the offset washer hole on the slab. At the mark, drill a ¾-inch hole 4 inches deep into the slab. Insert a lead shield into the hole, set the anchor and washer in position and screw a ½-inch lag bolt into the shield. Set a post support *(inset)* inside each anchor. Be sure that you can get at the bolthead under the support with a wrench; you will need to loosen the bolt and turn the washer to adjust the position of the post.

4 **Marking the corner posts for height.** Place the post temporarily in an anchor post and brace it plumb with 2-by-4s *(below)*; then tack a straight length of joist lumber to the end of the joist plate, level it and mark the post *(below, right)* along the bottom edge of the joist. Next, mark the proper length of the joist by reaching behind it and marking the outside edge along the post.

POST

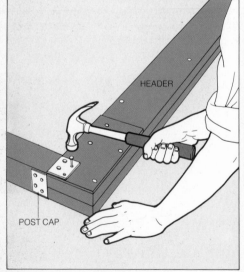

HEADER

POST CAP

5 **Assembling the posts and header.** Nail together two lengths of header lumber *(charts, page 107)*, install post caps at the ends and place ½-inch plywood behind the headers. Nail the corner posts into the valve of the caps *(above)*. Brace the posts in the anchors as in Step 4, and nail the anchor to the posts.

6 **Plumbing the new wall.** Remove the braces, plumb a corner post with a level and tack a 2-by-4 brace between the header and joist plate to hold the wall in alignment while the joists and rafters are installed. Plumb and secure the other corner post in the same way. Check the length between the existing wall and the headers to make sure it is the same on both sides.

7 **Installing the end joists.** With a helper, align the bottom of an end joist with the joist plate and the header, then butt-nail the joist to the plate. Recheck the corner post for plumb, level the joist and nail it to the header. Screw 3-inch angles into the corners at which the joist meets the plate and header. Install the other end joist in the same way and remove the braces.

8 **Hanging the other joists.** Nail metal hangers on 16-inch centers to the joist plate and the header, and drop the joists into the hangers. To prevent the header from bowing outward in the center when you nail the joists in place, have a helper push against the front of the header with a 2-by-4, so that the header butts tightly against the joist ends.

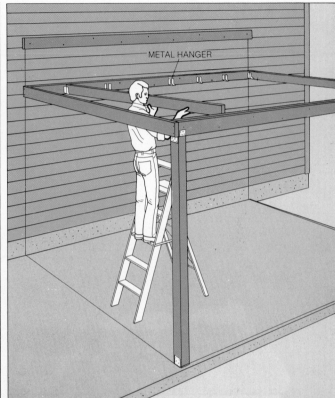

METAL HANGER

METAL HANGER

RAFTER ANCHOR

9 **Installing the rafters.** After marking and cutting the rafters (*pages 104-105*), nail rafter anchors to the top of the header and metal hangers to the rafter plate, both at 16-inch centers. Cut the hanger bottoms out so the rafters will rest firmly against the plate. Brace the end rafters with angles at the plate and header. Then nail the other rafters into the hangers and anchors. Nail ¾-inch trim to the rafter overhang and install sheathing and roofing materials.

The Gable Dormer—a Tiny House atop the Roof

A dormer that is built into a roof can bring sunlight and fresh air to a dim and stuffy attic. It can form the starting point for a new room, and thus add to the capacity and value of your home. And, in one type at least, it is something that you can build yourself.

A long shed dormer stretching the length of a roof needs professional planning. But constructing two or more gable dormers of the kind shown below does not call for the skill of an architect. The work consists of four jobs of carpentry: cutting an opening in the existing roof, fitting a gabled frame into the opening, installing a window in the front wall of the frame and covering the frame with roofing and siding.

Before tackling these jobs, check your attic space: the dormer ceiling should be at least 7 feet above the attic floor, and at least 2 feet below the main house ridge. Determine the size and spacing of the attic joists (your local building department needs this information to be sure the joists can bear a "live load"—usually 40 pounds per square foot).

Then plan the dormer itself in a detailed scale drawing on graph paper (Step 1, opposite, top), showing exactly where the roof opening will be cut—this drawing must accompany your application for a building permit—and draw up a shopping list. The front wall of the dormer should be high and wide enough to keep the window from looking cramped, and the pitch of the dormer should match that of the main roof. Choose a roofing material that harmonizes with the house roof—or better yet, build the dormers at a time when you are reroofing, so that all the roofing will be identical.

Erecting a dormer is mainly a matter of cutting and framing a fairly complex pattern of rafters and studs (below). Bird's-mouth cuts at the ends of rafters (page 104) are optional in this small, light structure, but many of the 2-by-4s must be cut at precise angles, laid out with a T bevel (page 73) and a steel framing square (page 105). Inside the dormer, paint or stain the top header if it is lower than the attic ceiling, and install ceiling joists as nailing surfaces for wallboard.

Use a prehung window with an exterior casing attached at the factory; look for one with sashes that raise high enough so the exterior panes can be easily cleaned from inside the dormer. Coat the window with a primer before setting it into the rough frame opening. If your window comes with a brace, leave it in place until the installation is completed.

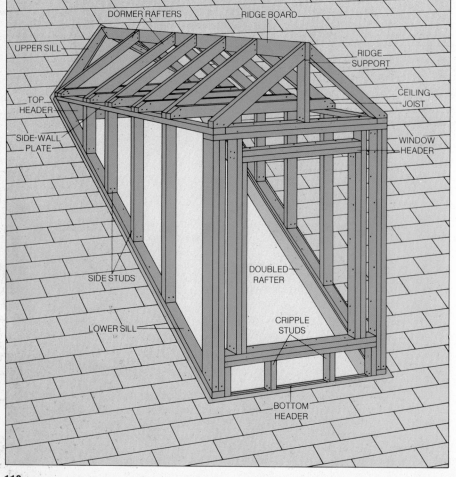

Anatomy of a dormer. The skeleton of a gable dormer consists entirely of 2-by-4s, except for a 2-by-6 ridge board. At the sides, this skeleton is supported by doubled rafters in the existing roof. The front wall is supported by a double bottom header of the same width and thickness as the roof rafters. A top header at the back of the dormer closes the roof opening. Upper and lower sills define the edges of the structure and provide nailing surfaces for flashing and siding; the side studs and front wall rise from the lower sill and the bottom header. Side-wall plates running from the front wall to the top of the lower sill support the dormer rafters. Cripple studs below the window opening, and ceiling joists below the dormer rafters, complete the frame.

Labels on diagram: DORMER RAFTERS, RIDGE BOARD, UPPER SILL, RIDGE SUPPORT, TOP HEADER, CEILING JOIST, SIDE-WALL PLATE, WINDOW HEADER, SIDE STUDS, DOUBLED RAFTER, LOWER SILL, CRIPPLE STUDS, BOTTOM HEADER

Framing the Dormer

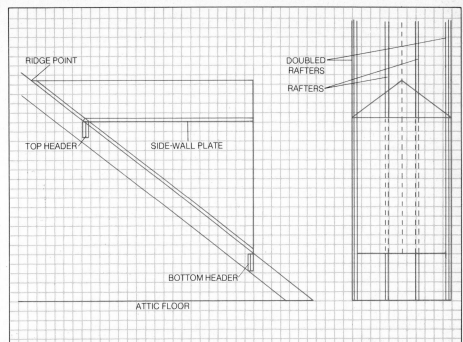

RIDGE POINT

DOUBLED RAFTERS

RAFTERS

TOP HEADER

SIDE-WALL PLATE

BOTTOM HEADER

ATTIC FLOOR

1 Planning the construction. On a sheet of graph paper, draw the profile *(right)* and front *(far right)* of the dormer to scale. Start with the profile of the roof, drawn to the correct slope *(page 66)*; for the front, start with the rafters and doubled rafters. Locate the lower edge of the top header at least 6 feet 8 inches above the attic floor, then draw in the profiles of a side-wall plate and bottom header at the correct heights: the bottom of the plate should be level with the upper edge of the top header, and the upper edge of the bottom header should be at least 2 feet above the floor. On the view of the dormer front, draw horizontal lines at these heights across the doubled rafters. Run a vertical dotted line up through the centers of these horizontals to locate the gable ridge, then draw the entire gable to the pitch of the house roof. On the profile, locate the point where the gable ridge meets the roof by drawing a horizontal line from the top of the gable to the line of the roof.

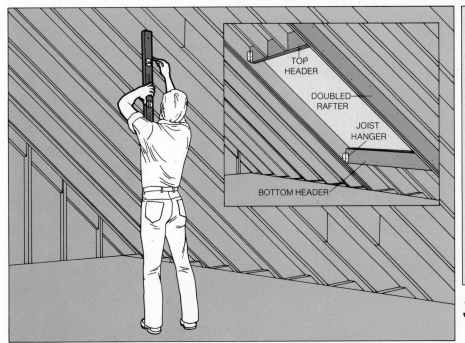

TOP HEADER

DOUBLED RAFTER

JOIST HANGER

BOTTOM HEADER

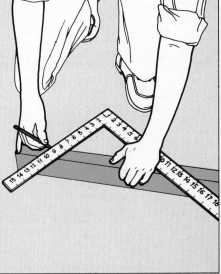

2 Making the opening. Using the graph-paper drawings as a guide, mark an outline of the roof opening on the undersides of the rafters; then, working with a level, draw plumb lines on the rafters to be cut. Double the rafters at the sides of the opening with lumber matching the dimensions of the rafters, using the methods shown on pages 72-75; nail these lengths to the outer sides of the rafters with tenpenny nails.

Drive eightpenny nails up through the sheathing at the corners of the outline of the roof opening. Outside the roof, snap chalk lines between

the nails to mark an outline of the opening. Cut through the roof along the outline by the method shown on page 117, Steps 1-3, cutting the rafters along the plumb lines. Outside the opening, make a rough outline of the dormer edges; include the ridge peak and allow space for the upper and lower sills, sheathing and siding. Remove all roofing material to clear the roof a minimum of 1 inch outside this outline. In the attic, nail four joist hangers, each 3 inches wide, to the doubled rafters at the corners of the opening, and install doubled top and bottom headers by the method shown on page 118, Steps 4-5.

3 Installing the lower sills. Set a 2-by-4 long enough for a lower sill on its narrow side and mark it for a plumb cut, using a steel framing square and the method shown on page 105, Step 1. Position the square with the 12 mark on the body, and the slope in inches (9 in this example) on the tongue along the same edge of the 2-by-4, and draw a line along the edge of the tongue. Cut the 2-by-4 along this line, then set it outside the roof opening, with the cut end flush to the front of the bottom header. Mark the other end for a square cut alongside the top header; make this cut, and fasten the 2-by-4 to the roof with 16-penny nails spaced 16 inches apart and driven into the doubled rafter below. Mark, cut and nail the other lower sill by the same method.

111

4 **Calculating the upper sill angles.** Outside the roof snap a vertical chalk line on the sheathing above the center of the roof opening; drive a nail loosely on the line at the point where the top of the dormer ridge board will intersect the house roof, and rest a 2-by-4 against this nail and the outer corner of a lower sill. Using a T bevel *(page 73)*, draw a line on the 2-by-4 at the angle where it meets the sill. Cut along this line, remove the temporary nail from the roof and set the sill in place. Mark the sill along its intersection with the chalk line and cut it at this angle. Repeat for the other upper sill and fasten both to the roof with nails driven into the rafters below.

5 **Assembling the front wall.** Frame and brace the front wall and its rough window opening at the same time to make the assembly shown below. For the wall, nail pairs of 2-by-4s together to form the posts and use a framing square to fit their bottom edge to the pitch of the lower sills; to determine the height of the posts—they should rise to a point 3 inches below the top of the side-wall plate—use your graph-paper plan. Nail the front-wall plate to the top of the posts.

For the rough window opening, frame a space 1 inch wider and 1½ inches higher than the jambs of your prehung window to allow for shimming when the window is installed *(Steps 11 and 12)*. Make the window sill of two 2-by-4s, nailed between the posts. Run full-length studs up from the sill to the front-wall plate, nail the doubled window header between the studs and nail jack studs between the sill and the window header. Cut two cripple studs to fit between the sill and the bottom header, and nail them to the posts. Check the entire front-wall assembly for squareness by taking its diagonal measurements—they should be equal—then nail a 1-by-3 diagonal brace across the face of the wall.

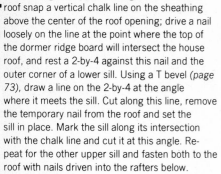

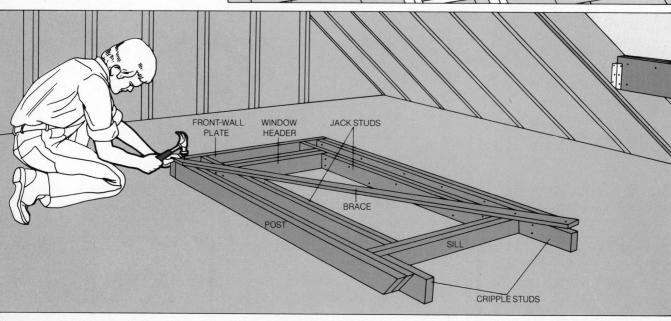

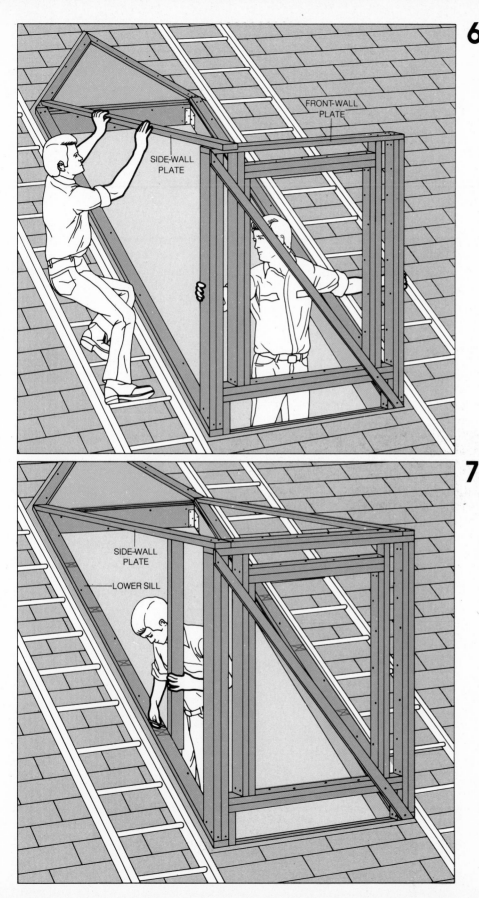

6 **Setting the side-wall plates.** Cut a 2-by-4 to fit the slope of the lower sill and, while a helper holds the front wall upright on the lower sills and the bottom header, set the side-wall plate in place. Nail the plate to the lower sill. With a level, plumb the front wall, then mark the side-wall plate at the point that lies over the front of the front-wall plate. Holding the side-wall plate in this position, nail it to the front-wall plate, then saw the side-wall plate flush to the front of the front-wall plate. Repeat for the other side-wall plate, and fasten a 2-by-4 between them with nails driven down into the front-wall plate.

7 **Installing the side studs.** Mark spaces for studs on 16-inch centers along the side-wall plate and, using a 2-by-4 and a level, transfer the marks to the lower sill; then hold the 2-by-4 between corresponding marks and draw a line along its edge for the cut at the lower sill. Make similar marks for all the side studs and cut the studs. Fasten the studs at the tops with 16-penny nails driven through the side-wall plates into the studs; at the bottoms, toenail the studs to the lower sills with eightpenny nails.

113

8 Fitting the ridge support. Cut the end of a 2-by-6 to fit the slope of the house roof and toenail it in place at the junction of the upper sills. While a helper supports the ridge board, set a 2-by-4 vertically at the center of the doubled front-wall plate, and mark the point at which it intersects the bottom of the ridge board. Cut the 2-by-4 to that height and toenail it to the ridge board and plate as a permanent ridge-board support.

9 Setting the rafters. Cut rafters as on page 105, Steps 1-3 (eliminate the bird's-mouths, if you wish, by extending the plumb cut at the side-wall plate). Set a rafter on the side-wall plate and, while a helper holds a level against the front wall and the rafter, mark the rafter's face on the ridge board. Nail the rafter to the ridge board, nail the other front rafter into place, then install the other rafters on 16-inch centers.

RIDGE SUPPORT
RIDGE BOARD
DOUBLED FRONT-WALL PLATE

SIDE-WALL PLATE
RIDGE BOARD

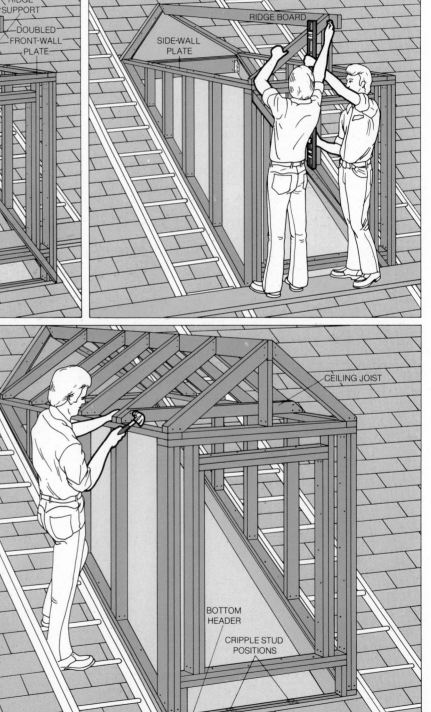

CEILING JOIST

BOTTOM HEADER

CRIPPLE STUD POSITIONS

10 Adding ceiling joists and cripple studs. Set a 2-by-4 on edge across the side-wall plates and against a pair of rafters; mark and cut it to fit both the slope and the plumb cut of the rafters, then nail it to the rafters. Install such a ceiling joist for each pair of rafters. Cut and install cripple studs to fit between the window sill and the bottom header at the points where cut roof rafters are nailed to the bottom header.

The framing of the dormer is now complete. Nail ½-inch plywood sheathing to its walls and roof by the methods shown on pages 27 and 69. At the eaves and gables put in 1-by-4 or 1-by-6 frieze boards to fit over the siding and under the roofing of the dormer.

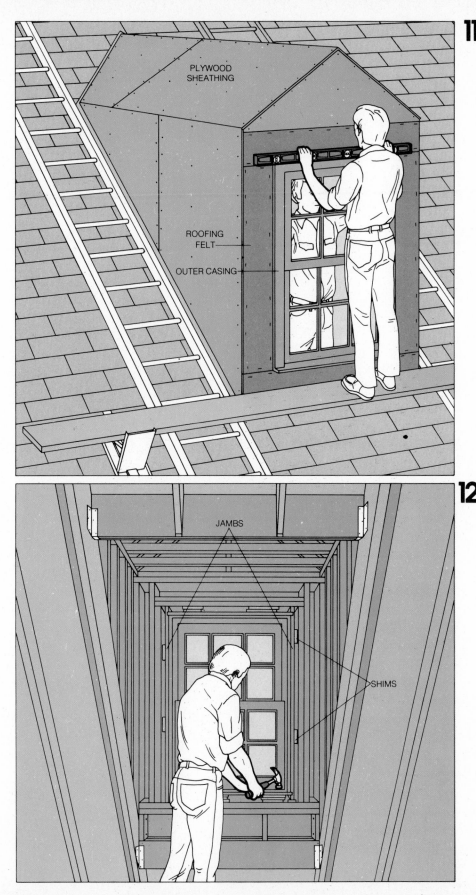

PLYWOOD SHEATHING

ROOFING FELT

OUTER CASING

JAMBS

SHIMS

11 **Leveling the window frame.** Before fixing the position of the window frame *(left)*, staple strips of 15-pound roofing felt around the outside edges of the window frame opening and prepare a drip cap *(page 123)*. Then, while a helper holds the top jamb of the window against the window header, check the outer casing with a level. One of the top corners will probably be slightly lower than the other: half sink an eightpenny casing nail through the casing and into the wall at the lower top corner, have the helper lower the other corner until it is level with the first, and half sink a nail at this corner. Check the sides for plumb and nail the bottom corners. Hammer in all the corner nails and drive additional nails along the casing at 12-inch intervals.

12 **Securing the window.** Drive at least two shims along the sides, the top and the bottom of the window jambs and beneath the side jambs—be sure that the shims do not shift the jambs out of plumb and level—then fasten the window with nails driven through the jambs and shims into the 2-by-4s *(left)*. Stuff fiberglass insulation into the cracks and beneath the sill.

Complete the exterior of the dormer by the methods shown elsewhere for flashing *(page 17)*, siding *(Chapter 2)* and roofing *(Chapter 3)*.

Cutting through a Roof for a Preframed Skylight

Many dramatic improvements can be made to your home by cutting through the roof. Perhaps the most striking is the addition of a skylight. A preframed plastic skylight makes it possible to brighten a dull room, add a sense of spaciousness to a cramped attic or create a greenhouse within your home at relatively low cost. Many, like the one sketched below, are built to open so they provide ventilation as well as light.

The plastic glazing is two panes thick for insulation. Since a skylight transmits three times as much light as a conventional window of the same size, the plastic can be ordered in glare-reducing tints as well as in clear. For extra control over glare and heat loss, you can hang a roller shade or mount an insulated plywood shutter *(page 121)* under the skylight.

Because the units come with frames attached, installing one is an easier job than you might expect it to be. Most are manufactured to fit between standard rafter spacing, so you can use existing rafters to support the sides of the skylight and to hold the headers that brace its ends. Locate the rafters and mark the position for the skylight from inside the house *(Step 1)*, then go outside to saw the opening. For most roofs, use the techniques on page 17 to take the roofing off before cutting the sheathing. If your roof is tile, slate or metal, or if you have a flat roof, you will need a contractor's help to prepare and finish the skylight opening. Then work indoors again to cut out the rafters within the hole and sup-

port the cut ends with double headers.

The boards or plywood sheets that sheathe the roof, beneath the roofing material, provide enough structural support to permit safe removal of sections of two or even three rafters. However, a skylight wider than 4 feet requires a professional's skill for installation.

After setting the skylight into the opening and sealing the edges, you can finish the inside of the well with wallboard *(page 121)*. A skylight located in an attic above a living area calls for a light shaft to the room below *(page 119, Steps 1-6)*, and the unit itself will include a removable, long-handled crank or pole so it can be opened and closed without a ladder. Remove the regular crank before sheathing the light shaft.

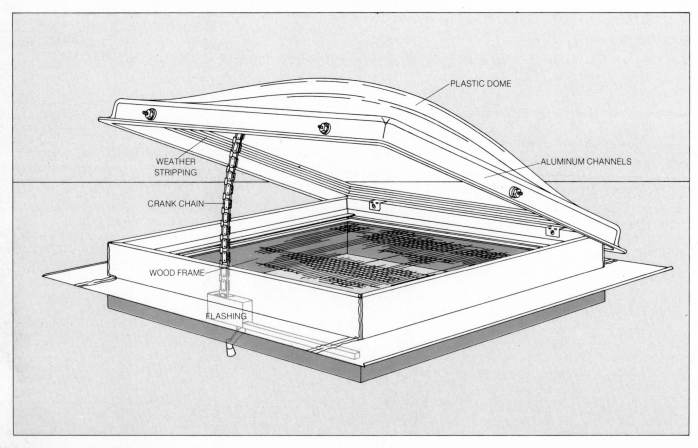

PLASTIC DOME

ALUMINUM CHANNELS

WEATHER STRIPPING

CRANK CHAIN

WOOD FRAME

FLASHING

Anatomy of a preframed skylight. Double plastic panes are tightly bonded to aluminum channels with weather stripping. The tops and sides of the wood frame are covered with flashing. A screen is built in, and the skylight is fitted with a crank to open and close it.

Installing the Skylight

1 Marking the hole. If rafters are exposed, mark four corners of the opening 3 inches longer than the skylight on the upper and lower rafter edges so that there is room for the headers. Make sure the opening is square by measuring diagonally between corners; the distances should be equal. Then from each corner mark, use a combination square to draw lines at right angles up the rafter sides to the roof sheathing. Drive an eightpenny nail through the sheathing at each corner. Go out on the roof and remove the roofing to expose at least 16 inches of felt on each side of the opening.

In a room with a finished ceiling, use a drill and wire probe to locate the rafters' inner edges. Draw a rectangle to fit between rafters as described above and cut a hole in the ceiling. Mark and drill holes through the roof and remove roofing as described above.

2 Cutting the roof. After checking skylight measurements against a rectangle drawn between the corner holes—the frame should fit inside the lines with spaces of about ¼ inch at each side and 3 inches at each end—use a utility knife to cut through the roofing felt. Then saw the hole with a saber saw or a circular saw set to a depth of about 1 inch.

3 Cutting rafters. Butting the ruler of a combination square against the rafter edge at the upper and lower ends of the opening, draw lines up each rafter in the opening. From inside, cut out the rafter sections with a handsaw.

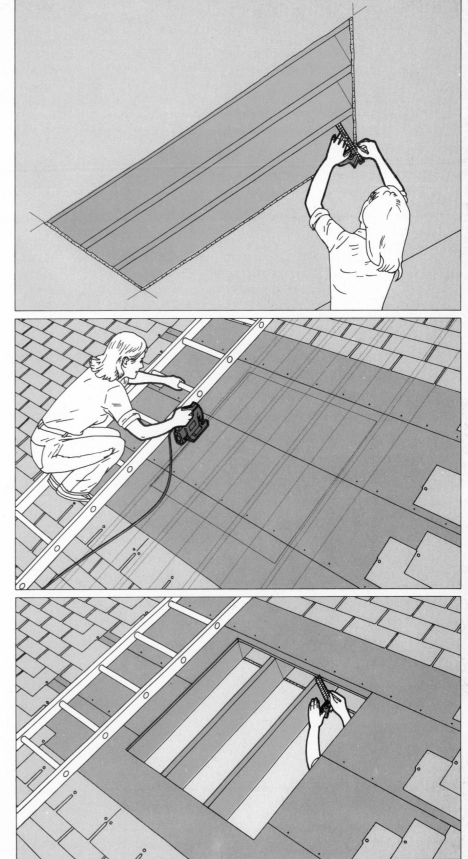

4 Attaching joist hangers. Using eightpenny nails, secure a 3-inch-wide hanger to each corner of the opening. Align the bottom of the hanger with the bottom of the uncut rafter and the outer side of the U-shaped support with the line drawn to the corner hole *(page 117, Step 3)*.

5 Installing headers. Cut four headers the width of the opening from lumber the same size as the rafters. Set one header into each facing pair of hangers and nail it to the cut rafter ends with 16-penny nails. Then fit a second header into each pair of hangers and attach it to the first header with nails driven in a staggered W pattern. Finally, nail the flanges of the joist hangers to the second headers.

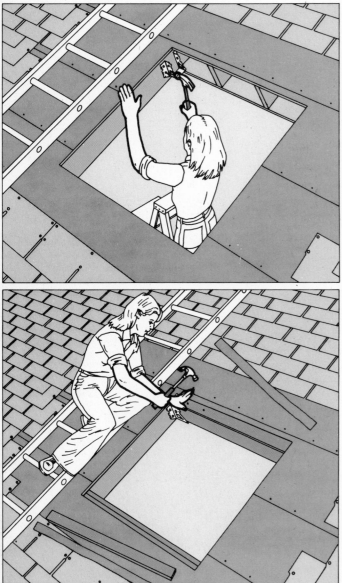

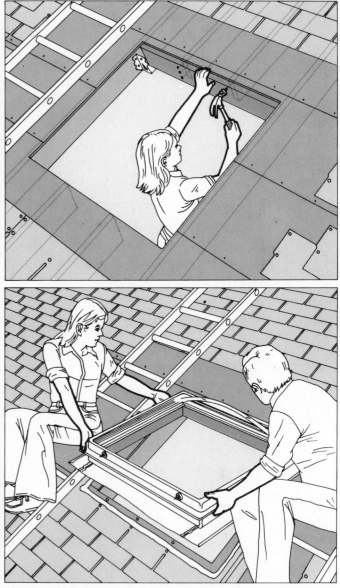

6 Covering the headers. Nail 3-inch-wide sheathing patches over the tops of the headers to make the opening level with the roof. Cut 5-inch strips of roofing felt and nail them over the headers so that they overlap the old felt. Replace the shingles on the bottom edge of the skylight.

7 Positioning the skylight. With the aid of a helper, set the skylight in the opening so that the top and side flanges of the flashing cover the felt borders but the bottom flange lies directly over the newly installed shingles. Drive 1-inch roofing nails into the upper flange at 12-inch intervals.

Secure the ''slater's edges'' of the flashing along the sides with cleats *(page 17)*. Then replace the roofing over the top and side flashing, keeping the nails beyond the edges of the flashing.

8 **Finishing the skylight.** Use eightpenny finishing nails to secure the wood frame of the skylight to the headers and rafters. If the headers and rafters project below the skylight frame, rip 1-inch lumber to extend the bottom of the frame.

If the ceiling is finished, cover the bottom and inside surfaces of the frame with wallboard or ¾-inch plywood. Cover the headers with wallboard and enclose the corners with metal corner bead or wood trim.

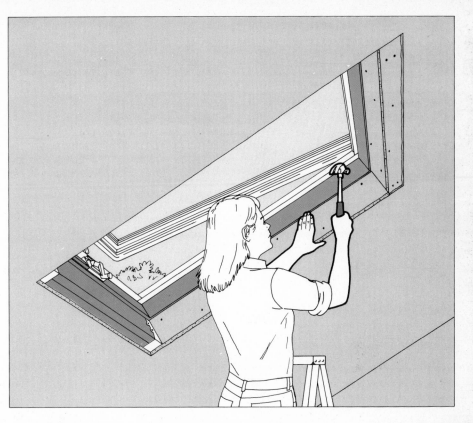

Bringing Light to a Room below the Attic

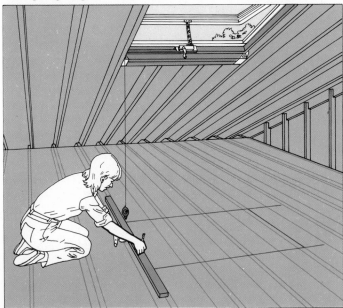

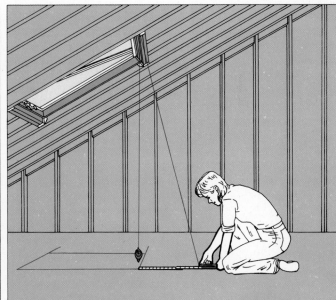

1 **Planning the shaft.** Drop a plumb line from each corner of the skylight and mark the floor. Then draw a rectangle connecting the marks. If the rafters and joists are offset as shown, the inner edge of one joist will lie inside the corners of the skylight. Use a straightedge to extend that side of the rectangle outward to the next joist. Where the joists are not exposed, use a drill and wire probe to locate their inner edges.

2 **Enlarging the shaft.** To angle the ends of the shaft beyond the ends of the skylight, tack a string to one corner of the skylight frame and pull the string taut to the floor at the desired angle. Mark the floor, measure from the rectangle to the mark and make a mark at this distance on the other side of the line. If your shaft is to be angled at both ends, repeat this procedure on the other side, then redraw the rectangle.

119

3 **Opening the ceiling.** If the ceiling joists are not exposed, cut the floor out along the lines you have drawn with a circular saw. Use a combination square to draw a line down the joists to the ceiling below at each corner of the opening. Then drill a hole in the ceiling where each line hits the ceiling. Working from below, draw a rectangle to connect the holes in the ceiling. Cut the ceiling out.

4 **Cutting the joists.** Brace the ceiling with 2-by-4 T jacks located 12 inches away from both ends of the opening. For the jack tops, use 2-by-4s longer than the opening is wide at the ceiling; jam them tightly in place with 2-by-4 props. Shim the props so they fit snugly. Then, working from above, use the combination square to draw lines at a 90° angle to the floor at both ends of the joists within the opening. Saw off the joists sections and follow the techniques on page 118, Steps 4-6 to install headers.

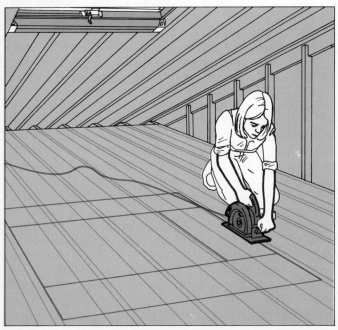

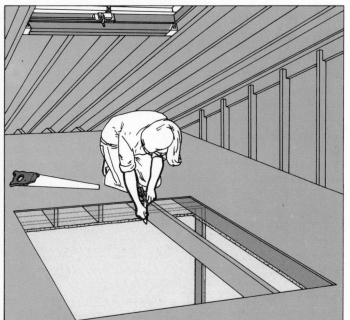

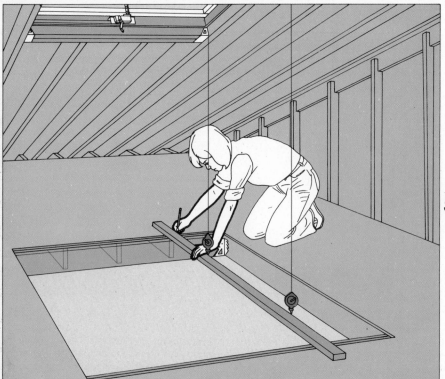

5 **Framing the opening.** On the side of the opening where you cut out an extra joist section as explained on page 117, Step 1, drop two plumb lines from the corners of the skylight. Line up a straightedge with the plumb bobs. Mark the position of the straightedge on both headers. Then cut a piece of joist lumber to fit between the headers and install it with joist hangers at the mark. Attach a 1-by-2 nailing strip to the bottom inner corner of the uncut joist at the side of the opening. On the opposite side, nail a piece of joist lumber alongside the existing joist.

Replace the flooring, if any, so that it extends to the edges of the shaft opening.

6 **Finishing the shaft.** Install a 2-by-3 sole plate around the four edges of the opening, and, using 2-by-3s spaced 16 inches on center, frame out the edges of the shaft. Determine the bevel cut for any necessary angle by placing a 2-by-3 against the sole plate and rafter and scribing the end of the board *(left)*.

Staple insulation to the studs so that the vapor barrier faces toward the center of the shaft. Cut wallboard to size and nail it inside the shaft and under the exposed headers and joists in the ceiling below. Use joint compound and tape to seal corners and joints, then enclose corners with metal corner bead or wood trim. Sand the compound smooth and, if desired, paint the inside of the light shaft a bright color for maximum light reflection.

Sliding Panels that Seal the Light Shaft

Insulating shutters. To shut the skylight partially—or completely—mount sliding plywood panels backed with fireproof, rigid insulation in 1½-inch aluminum channels. Cut two pieces of channel each twice the length of the skylight opening. Center the channels on either side of the opening so the ends extend an equal distance beyond either end. Drill holes in the channel flanges so you can screw them to the ceiling.

Cut ½-inch plywood the length of the skylight to fit between the tracks. Glue an insulation panel to the plywood. After the glue dries, cut the insulated plywood piece in half and slide both halves into the channels, insulation facing upward. Install a ¾-inch finger pull in the center of both panels so that you can use your finger or a pole to slide the panels back and forth.

Portable Platforms for a Rooftop Deck

A flat roof over an attached garage or a single-story extension of the house can easily be converted into a sun deck. All that is necessary is to install a door onto the roof and to build a row of narrow platforms. Fastened to each other and anchored to the house by a railing, the sections can be taken apart and moved individually should it ever become necessary to clean or repair the roof.

By way of preparation, first make sure that the roof is pitched no more than 1 in 12 or you will not find joist lumber wide enough for more than a shallow deck. Check that the rafters can carry the extra load of about 40 pounds per square foot imposed by the deck and the people using it. For example, 2-by-6 rafters spaced 16 inches apart will support the deck if

their ends rest on bearing walls no more than 9 feet apart. Rafters of 2-by-8s can span as much as 11 feet 8 inches.

If the rafters are strong enough, plan to replace a window overlooking the roof with a door. The window opening should be at least 32 inches wide with the top at least 6 feet 2 inches above the doorsill; the sill for the new door must be at least 5 inches above the roof so the deck will be level with or below it. To achieve that, you may have to install a raised doorsill. Additionally, the space below the window must be clear of pipes and wiring. Since a deck is often considered an addition to a house, you may have to submit a sketch of your plans to local authorities and secure a building permit.

For materials, you need a prehung ex-

terior door to fit the opening in the wall, and molding for the interior casing. The deck—except for 1-by-4 middle railings—is made of 2-inch boards pressure-treated with preservative to keep them from rotting. Each platform section requires three joists wide enough to compensate for roof slope *(page 124)* and 2-by-6 lumber for floor planks. Railing posts and top railing are made of 2-by-4s.

To allow space for railing supports, plan your deck to extend no closer than 3 inches from the roof edge. The job will look best if all platform sections are the same width—between 2½ and 4 feet, depending on the size of your roof. Begin the job by installing the door so that you have easy access to the roof while you assemble the deck.

Putting in a Door

1 Making the opening. To enlarge a window into a door, remove the window as well as the wallboard, the drip cap above the window, insulation and framing below the opening. If the siding is wood, set a circular saw to its thickness and cut through below the window to the bottom of the sole plate at the base of the wall; pry the pieces loose. Reset the saw to the sheathing thickness, then guide it along the jack studs and the bottom of the sole plate to complete the door opening. Finally cut the sole plate from between the jack studs.

Cut brick veneer in the same way, using a masonry-cutting blade in the saw.

2 **A drip cap for the doorway.** Wearing gloves, cut and bend a strip of 6-inch flashing metal to fit across the opening. Use a 2-by-4 to make the two 90° bends needed, placing the strips lengthwise over the board edge. Make one bend ½ inch from one edge, the second 1½ inches from the first and in the opposite direction. The remaining 4-inch flat section goes under the siding at the top of the opening.

SIDE JAMB

TOP CASING

SIDE CASING

SIDE JAMB

3 **Installing the doorframe.** Remove the door from the frame, leaving any shipping braces in place, and install it with shims so that the sill is level and more than 3½ inches above the roof. If the bottom of the opening gives that height, push the frame in, check which end of the sill is higher, and 6 inches above the sill drive a finishing nail through the side jamb into the jack stud. Level the sill with a shim under the low end and nail the other side jamb.

Fasten both jambs to the jack studs, nailing every 6 inches. Remove shipping braces.

If the bottom of the door opening is less than 3½ inches above the roof, use a combination of 2-by-4s and 1-by-4s to elevate the sill. Nail the exterior top and side casings to the header and jack studs behind. Fit the drip cap to the top casing and caulk the gaps between casings and siding, then hang the door.

4 **Finishing the doorframe.** The outside of the frame comes finished, but the inside needs molding, called casing, which is applied to the top and sides. Cut the top casing so that the lower edge is ¼ inch longer than the inside of the door-jamb (*inset*). Using eightpenny finishing nails, fasten the molding to the jamb, then to the header inside the wall, so that the lower edge is ⅛ inch above the inner edge of the top jamb. For a side casing, stand a molding strip on the floor, against a side jamb. Mark the strip with the direction of a miter cut where the inside edge meets the top casing. Repeat the procedure for the other side casing, miter the two strips and attach them like the top casing.

Assembling the Deck

1 **Measuring joist slope.** If the roof is pitched more than ¼ inch per foot, you will have to compensate for the slope in constructing the deck sections. To do so, lay a long 2-by-4 at a right angle from the wall of the house to the edge of the roof. Place a level on the board and lift one end to level it. Three inches from the edge of the roof, measure the distance between the roof and the bottom of the board.

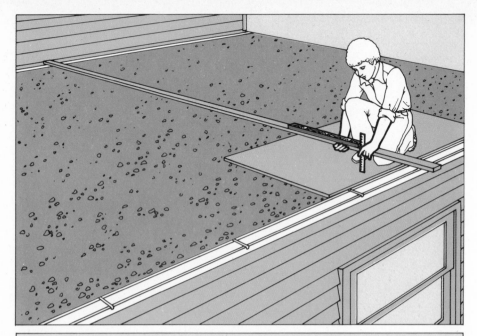

2 **Cutting the joists.** The measurement made in Step 1 enables you to snap a chalk line for angling the joists. To accommodate this long angled cut, you may need wide joist lumber. After trimming all joists to a length 3 inches less than roof width, mark one end of each 2 inches up from an edge; mark the other end 2 inches up plus the measurement taken in Step 1. Snap a chalk line between the marks and rip along it.

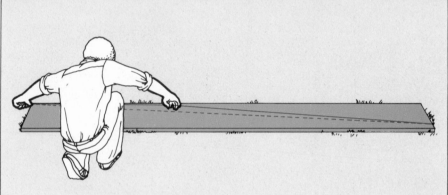

3 **Constructing the frames.** For each platform, fasten together three joists with 2-by-6s at both ends *(inset),* working at ground level. Use tenpenny galvanized finishing nails. Carry the platform frames to the roof and position them, clearing away gravel from beneath the joists. Nail on 2-by-6 decking, using a strip of ¼-inch plywood to space the planks. Countersink nails and shim low platforms to make an even surface.

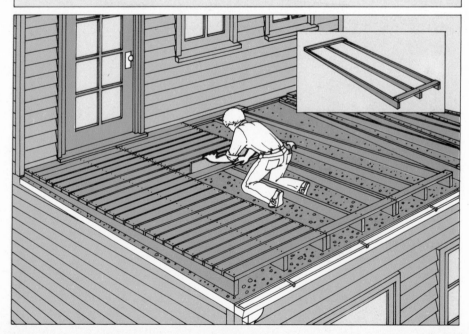

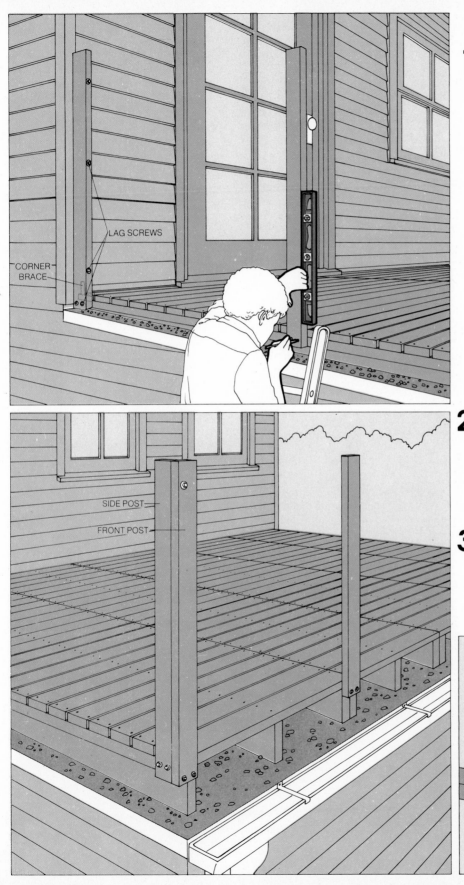

Attaching the Railing

1 **Erecting the side posts.** Cut 43-inch 2-by-4s for railing posts and then fasten them no more than 3 feet apart along the sides of the deck and at every platform joint across the front, using two 4-inch lag screws in each. Secure each with a 3-inch corner brace screwed inside the post to the deck. Anchor the two posts next to the house with three 6-inch lag screws into the siding.

LAG SCREWS

CORNER BRACE

2 **Turning the corner.** At the corners, screw the side post flush with the end of the joist. Make the front post overlap the side post and screw it both to the side post, at top and bottom, and to the end of the joist.

SIDE POST

FRONT POST

3 **Adding the top rail.** Attach 2-by-4s to the post tops with 3-inch No. 12 galvanized wood screws so the railing overhangs the posts equally on both sides. If you join boards to make a long rail, butt them on top of a post. Screw 1-by-4 middle railings horizontally across posts, spacing them vertically about 4 inches.

Picture Credits

The sources for the illustrations in this book are shown below. Credits for the pictures from left to right are separated by semicolons, from top to bottom by dashes.

Cover—Fred Maroon. 6—Fred Maroon. 8 through 11—Drawings by Vantage Art, Inc. 12 through 21—Drawings by Fred Bigio from B-C Graphics. 22—Fred Maroon. 24—Drawing by John Massey. 26 through 29—Drawings by Ray Skibinski. 30 through 36—Drawings by Peter McGinn. 37—Drawing by Walter Hilmers Jr. 38 through 41—Drawings by Gerry Gallagher. 42 through 47—Drawings by Vicki Vebell. 48 through 53—Drawings by Peter McGinn. 54 through 59—Drawings by John Massey. 60—Drawings by John Massey; Portland Cement Association (2). 61—Drawings by John Massey; Portland Cement Association (2). 62,63—Drawings by Nick Fasciano. 64A—Robert Lautman, William Turnbull Jr., Architect. 64B—John Fulker, Barton Myers, Architect. 64C—Donald Singer, Photographer and Architect—Bird & Son, Inc. 64D—Pierre Botschi and Derek Walker, designed and developed by Milton Keynes Special Projects Group, Architect in charge, Pierre Botschi—Valerie Batorewicz, Photographer, Architect, Inventor and Builder. 64E—Fred Ward from Black Star, Steve Baer, Architect, Zomeworks Corporation—John Zimmerman, David C. Harrison, Inventor, Zomeworks Corporation, Crystal Creek Construction Company, Designer and Contractor (2). 64F—Creative Photographic Services, William Morgan Architects. 64G—Shakertown Corporation, Ronald E. Thompson, Architect—Emmett Bright, Peter Schneck, Architect. 64H—Fred Maroon. 66—Drawing by John Massey. 68 through 75—Drawings by Nick Fasciano. 76 through 81—Drawings by Vicki Vebell. 82 through 85—Drawings by Walter Hilmers Jr. 86 through 90—Drawings by John Massey. 91—Courtesy Library of Congress. 92 through 97—Drawings by Peter McGinn. 98 through 101—Drawings by Walter Hilmers Jr. 102—Fred Maroon. 104 through 115—Drawings by Fred Bigio from B-C Graphics. 116 through 121—Drawings by Gerry Gallagher. 122 through 125—Drawings by Ray Skibinski.

The following persons also assisted in the making of this volume by preparing the preliminary sketches from which the final illustrations were drawn: Roger C. Essley, Fred Holz, Joan S. McGurren, W. F. McWilliam.

Acknowledgments

The index/glossary for this book was prepared by Mel Ingber. The editors also wish to thank the following: Alexandria Building Dept., Alexandria, Va.; The Aluminum Assn., New York, N.Y.; Karl A. Baer, National Housing Center Library, Washington, D.C.; Paul Bechtold, Venturama Skylight Corp., Port Washington, N.Y.; Alan Black, Charlotte, N.C.; Myron Blakesley, K-Lath Division, Tree Island Steel, Inc., Monrovia, Calif.; Thomas J. Boyd, Follansbee, W. Va.; Mr. and Mrs. Harold Brown, Burke, Va.; California Redwood Assn., San Francisco, Calif.; Charles Campbell, William R. Garrison Jr., Robert Johnson, Samuel B. Maize Jr. and Charles Wood, Virginia Roofing Corp., Alexandria, Va.; John Colline, Portland Cement Assn., Arlington, Va.; Dan Crawford, Stancliff Hardware, New Carrollton, Md.; Richard A. Dickson, Kaiser Aluminum and Chemical Corporation, Oakland, Calif.; Patrick H. Donahue, Bird and Son, Inc., East Walpole, Mass.; Lelland L. Gallup, Cornell University, Ithaca, N.Y.; Glenn Halme, American Plywood Assn., Tacoma, Wash.; Patricia Harms and Frank Randall, Portland Cement Assn., Skokie, Ill.; Joe Howell and Gerald Miller, Monroe Development Corp., Potomac, Md.; Milton Jernigan Jr., Jernigan's AAA Rental Centers, Lanham, Md.; Johns-Manville, Denver, Col.; Harris Kenner, M. W. Dunton Co., Providence, R.I.; Judy McCullough, O. Ames Company, Parkersburg, W. Va.; Donald Millikin, Cedar Roofs Inc., McLean, Va.; George B. Mueller, Architectural Engineering Products Co., Richton Park, Ill.; Frank Obal, Council of Forest Industries of British Columbia, Vancouver, B.C.; Frank Papso, Kaiser Aluminum, Sylvania, Ohio; Dean C. Patterson, Brick Institute of America, McLean, Va.; Virgil G. Peterson, Red Cedar Shingle and Handsplit Shake Bureau, Bellevue, Wash.; Hildegard Popper, Asphalt Roofing Manufacturers' Association, New York City; Thomas S. Rimel and Seth L. Warfield, Jack's Roofing Co., Inc., Washington, D.C.; Lee Roxborough, Mt. Vernon Roofing and Siding Co., Alexandria, Va.; Dr. Theodore A. Sande, National Trust for Historic Preservation, Washington, D.C.; Norval (Happy) Sipes, Wheaton, Md.; Robert Smith, Kettler Brothers, Inc., Gaithersburg, Md.; Gilbert Wolff, National Plastering Institute's Joint Apprenticeship Trust, Washington, D.C. *The following persons also assisted in the writing of this book:* Lewis H. Diuguid, William Garvey, Tom Levno, William R. Loch, Susan K. Nelson, Randy Rabin and Peter Rohrbach.

Index/Glossary